M336
Mathematics and Computing: a third-level course

GROUPS & GEOMETRY

UNIT GE5
GROUPS AND SOLIDS IN THREE DIMENSIONS

Prepared for the course team by
David Crowe

The Open University

This text forms part of an Open University third-level course.
The main printed materials for this course are as follows.

Block 1
Unit IB1   Tilings
Unit IB2   Groups: properties and examples
Unit IB3   Frieze patterns
Unit IB4   Groups: axioms and their consequences

Block 2
Unit GR1   Properties of the integers
Unit GR2   Abelian and cyclic groups
Unit GE1   Counting with groups
Unit GE2   Periodic and transitive tilings

Block 3
Unit GR3   Decomposition of Abelian groups
Unit GR4   Finite groups 1
Unit GE3   Two-dimensional lattices
Unit GE4   Wallpaper patterns

Block 4
Unit GR5   Sylow's theorems
Unit GR6   Finite groups 2
Unit GE5   Groups and solids in three dimensions
Unit GE6   Three-dimensional lattices and polyhedra

The course was produced by the following team:

Andrew Adamyk (BBC Producer)
David Asche (Author, Software and Video)
Jenny Chalmers (Publishing Editor)
Bob Coates (Author)
Sarah Crompton (Graphic Designer)
David Crowe (Author and Video)
Margaret Crowe (Course Manager)
Alison George (Graphic Artist)
Derek Goldrei (Groups Exercises and Assessment)
Fred Holroyd (Chair, Author, Video and Academic Editor)
Jack Koumi (BBC Producer)
Tim Lister (Geometry Exercises and Assessment)
Roger Lowry (Publishing Editor)
Bob Margolis (Author)
Roy Nelson (Author and Video)
Joe Rooney (Author and Video)
Peter Strain-Clark (Author and Video)
Pip Surgey (BBC Producer)

With valuable assistance from:

Maths Faculty Course Materials Production Unit
Christine Bestavachvili (Video Presenter)
Ian Brodie (Reader)
Andrew Brown (Reader)
Judith Daniels (Video Presenter)
Kathleen Gilmartin (Video Presenter)
Liz Scott (Reader)
Heidi Wilson (Reader)
Robin Wilson (Reader)

The external assessor was:

Norman Biggs (Professor of Mathematics, LSE)

The Open University, Walton Hall, Milton Keynes, MK7 6AA.

First published 1994. Reprinted 1997, 2002, 2005, 2009.

Copyright © 1994 The Open University

All rights reserved. No part of this publication may be reproduced, stored in a retrieval system or transmitted in any form or by any means, without written permission from the publisher or a licence from the Copyright Licensing Agency Limited. Details of such licences (for reprographic reproduction) may be obtained from the Copyright Licensing Agency Ltd of 90 Tottenham Court Road, London, W1P 9HE.

Edited, designed and typeset by the Open University using the Open University TeX System.

Printed in Malta by Gutenberg Press Limited.

ISBN 0 7492 2173 9

This text forms part of an Open University Third Level Course. If you would like a copy of *Studying with the Open University*, please write to the Central Enquiry Service, PO Box 200, The Open University, Walton Hall, Milton Keynes, MK7 6YZ. If you have not already enrolled on the Course and would like to buy this or other Open University material, please write to Open University Educational Enterprises Ltd, 12 Cofferidge Close, Stony Stratford, Milton Keynes, MK11 1BY, United Kingdom.

1.3

The paper used for this book is FSC-certified and totally chlorine-free. FSC (the Forest Stewardship Council) is an international network to promote responsible management of the world's forests.

CONTENTS

Study guide	**4**
Introduction	**5**
1 Isometry groups in two dimensions	**6**
1.1 Cyclic and dihedral groups	6
1.2 The groups $O(2)$ and $SO(2)$	8
2 Isometries in three dimensions	**11**
2.1 Isometries in $\mathbb{R}^3$	11
2.2 Finite groups of isometries in $\mathbb{R}^3$	13
2.3 Reflections of $\mathbb{R}^2$ can be rotations in $\mathbb{R}^3$	18
3 Direct groups	**19**
4 Finite subgroups of $O(3)$	**21**
4.1 The coset decomposition revisited	21
4.2 The dihedron with n equilateral vertices	22
4.3 The even and odd cases	23
4.4 The tetrahedron	26
4.5 The cube and octahedron	27
4.6 The dodecahedron and icosahedron	27
5 A complete catalogue	**28**
5.1 Finite subgroups of $SO(3)$	28
5.2 The remaining finite subgroups of $O(3)$	33
5.3 A complete catalogue	35
Solutions to the exercises	**42**
Objectives	**45**
Index	**46**

STUDY GUIDE

The first four sections are of roughly the same length. Section 5 is longer; as we note, some of it may be skipped if you are short of time.

Section 1 revises and consolidates work from earlier in the course; much of the material should be familiar so you should not find this section hard.

Section 2 extends these ideas to deal with three-dimensional space.

In Section 3 we calculate the rotation groups of some familiar figures. This section is closely tied to video programme *VC4A*, which you should watch after working through the printed material in Section 3. The video notes provide work for you to do before you watch the video programme and after you have seen it.

Section 4 deals with the indirect symmetries of the figures you saw in the video programme, and others. The section gives a proof of an important theorem which enables these groups to be completely classified.

Finally, in Section 5 we see that there are essentially no other finite symmetry groups in $\mathbb{R}^3$. The proof of this fact is moderately long, and if you are short of time you may wish to skip it. (That is, you may go straight to the statement of Theorem 5.3 on page 33.) However, you should understand the statement of Theorem 5.4 in Subsection 5.2, and you should study the complete catalogue in Subsection 5.3 carefully.

There is no audio programme for this unit.

Apart from an occasional use of the Isometry Toolkit, you do not need the *Geometry Envelope* for this unit.

INTRODUCTION

You have already met a wide variety of symmetry groups associated with patterns and figures in the plane $\mathbb{R}^2$. Many of these groups are infinite. For example, the symmetry group of a point lattice contains all the translations of the (infinite) lattice, and thus has infinitely many elements. Some finite figures also have infinite symmetry groups! Consider a circle, for example. Clearly, rotation through any angle about the centre is a symmetry of this figure, and there are certainly infinitely many of these.

Throughout this unit we shall concentrate on *finite* symmetry groups, and in particular we shall not spend much time on translations.

In *Unit IB2* you saw that the notion of symmetry group can be formalized in terms of isometries — distance-preserving maps from the plane to itself. In particular, we saw that any element of a finite group of plane isometries is either a rotation or a reflection. The main aim of this unit is to extend our theory to the more interesting problem of symmetry in three-dimensional space $\mathbb{R}^3$.

Since classical times it has been known that certain (finite) solid objects enjoy remarkable symmetry properties. The best known examples are the Platonic (or regular) solids, shown in Figure 0.1.

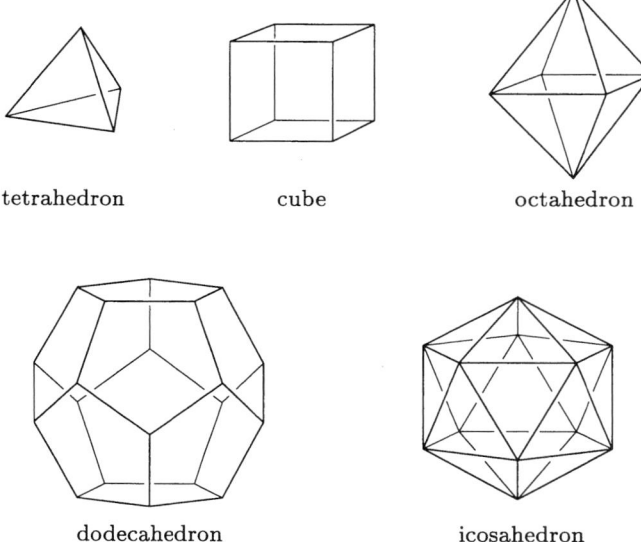

Figure 0.1 The Platonic solids

By the end of this unit you should understand quite a lot about the symmetries of these objects, and also have a good idea why there are *only* these five regular solids.

In Section 1 we consolidate certain results from earlier in the course, involving finite symmetry groups of the plane, and we establish some important theorems, which are used throughout the remainder of the unit. In Section 2 we show how these ideas can be extended to apply in three-dimensional space and attempt to give you a 'feel' for three-dimensional isometries.

Section 3, which is associated with the video programme, discusses the five regular solids in detail and classifies their groups of rotational symmetry. Part of the aim of this section is to assist your visualization of three-dimensional shapes. In Section 4 we study the symmetries which entail 'turning the object inside out' — in other words using a reflection of $\mathbb{R}^3$. Finally, in Section 5 we show that our search is complete, in the sense that all possible finite symmetry groups of $\mathbb{R}^3$ have been encountered.

1 ISOMETRY GROUPS IN TWO DIMENSIONS

1.1 Cyclic and dihedral groups

Throughout this course, from *Unit IB2* onwards, you have encountered the cyclic and dihedral groups, C_n and D_n ($n \in \mathbb{N}$). Not only are they important in their own right, but they are also essential in analysing the infinite symmetry groups of the two-dimensional lattices and wallpaper patterns (as you saw in *Units GE3* and *GE4*).

C_n and D_n are given in the Appendix to *Unit IB4*.

The reason for their importance is that they are the *only* groups of two-dimensional isometries that are of finite order. This fact has been lurking in the background throughout the Geometry stream of the course, so it may come as a shock to you to realize that, up till now in the course, you have not seen it proved! We shall now remedy this. .

The first step is to verify that any finite group of order $n > 1$ that consists solely of rotations about the origin (including, of course, the trivial rotation e) is in fact generated by $r[2\pi/n]$.

Exercise 1.1

Let G be a group of order $n > 1$, consisting of rotations about the origin. Let its elements be listed as

$$G = \{r[\theta_1], r[\theta_2], \ldots, r[\theta_{n-1}], e\},$$

where $0 < \theta_1 < \theta_2 < \cdots < \theta_{n-1} < 2\pi$, and e is regarded as $r[\theta_n]$ where $\theta_n = 2\pi$.

(a) Use the fact that G is a group to show that

$$\theta_{j+1} - \theta_j = \theta_1 \quad (j = 1, 2, \ldots, n-1).$$

(b) Deduce that $G = \langle r[\theta_1] \rangle$ and that $\theta_1 = 2\pi/n$.

Next, we need the result that for any *finite* group G of plane isometries, there is some point in the plane which is fixed by every element of G.

In fact, the easiest way to prove this uses a technique, which works for any number of dimensions, based on *group actions*, a concept which you have seen in several earlier units.

We write the effect of an element $g \in G$ on an element $x \in X$ as $g \wedge x$. Such an action satisfies the following three properties.

See Lemma 5.1 of *Unit IB2*.

Properties of a group action

(a) $g \wedge x \in X$, for all $g \in G, x \in X$;
(b) $e \wedge x = x$, for all $x \in X$;
(c) $(gh) \wedge x = g \wedge (h \wedge x)$, for all $g, h \in G$ and $x \in X$.

We can now prove the following theorem.

> **Theorem 1.1 Fixed point theorem**
>
> Let G be a finite group of isometries of the Euclidean space $\mathbb{R}^m$. Then there is at least one point of $\mathbb{R}^m$ which is fixed by all the elements of G.

Proof

Since the given symmetry group G is finite, we can list all its elements explicitly as $\{g_1, g_2, g_3, \ldots, g_n\}$, where $n = |G|$, the order of G. Now take any point of $\mathbb{R}^m$, which we shall call $\mathbf{a}$. The group action of G on $\mathbb{R}^m$ gives a set of n points $\{g_1 \wedge \mathbf{a}, g_2 \wedge \mathbf{a}, \ldots, g_n \wedge \mathbf{a}\}$ — this is the *orbit* of $\mathbf{a}$ under the action of G. One of these points will be $\mathbf{a}$ itself, since G contains the identity element which does not move anything. The crucial fact is that the *centroid* of all these points is fixed by every element of G.

To see this, note that the centroid of the above set of points is

$$\overline{\mathbf{a}} = \frac{1}{n}(g_1 \wedge \mathbf{a} + g_2 \wedge \mathbf{a} + \cdots + g_n \wedge \mathbf{a})$$

and consider the action of an arbitrary element g_r of G on $\overline{\mathbf{a}}$. Now g_r can be written as an orthogonal matrix transformation followed by a translation:

$$g_r = t[\mathbf{p}_r]\lambda[\mathbf{A}_r].$$

We have

$$g_r \wedge \overline{\mathbf{a}} = \mathbf{A}_r \left\{ \frac{1}{n}(g_1 \wedge \mathbf{a} + g_2 \wedge \mathbf{a} + \cdots + g_n \wedge \mathbf{a}) \right\} + \mathbf{p}_r.$$

Since $\mathbf{A}_r$ is a linear transformation, the term inside the curly brackets can be expanded, giving

$$g_r \wedge \overline{\mathbf{a}} = \frac{1}{n}\{\mathbf{A}_r(g_1 \wedge \mathbf{a}) + \mathbf{A}_r(g_2 \wedge \mathbf{a}) + \cdots + \mathbf{A}_r(g_n \wedge \mathbf{a})\} + \mathbf{p}_r.$$

There are n terms inside the curly brackets, and we now divide the $\mathbf{p}_r$ outside the brackets into n equal parts, allocating one part to each term on the inside. Thus

$$g_r \wedge \overline{\mathbf{a}} = \frac{1}{n}\{[\mathbf{A}_r(g_1 \wedge \mathbf{a}) + \mathbf{p}_r] + [\mathbf{A}_r(g_2 \wedge \mathbf{a}) + \mathbf{p}_r] + \cdots + [\mathbf{A}_r(g_n \wedge \mathbf{a}) + \mathbf{p}_r]\},$$

which means that

$$g_r \wedge \overline{\mathbf{a}} = \frac{1}{n}\{g_r \wedge (g_1 \wedge \mathbf{a}) + g_r \wedge (g_2 \wedge \mathbf{a}) + \cdots + g_r \wedge (g_n \wedge \mathbf{a})\}.$$

This can equally well be written as

$$g_r \wedge \overline{\mathbf{a}} = \frac{1}{n}\{(g_r g_1) \wedge \mathbf{a} + (g_r g_2) \wedge \mathbf{a} + \cdots + (g_r g_n) \wedge \mathbf{a}\}. \tag{1.1}$$

Now observe that the set $\{g_r g_1, g_r g_2, \ldots, g_r g_n\}$ is another listing of the elements of G, since it is a row of its Cayley table. So, in Equation 1.1, the term inside curly brackets is simply a re-ordering of the expression for $n\overline{\mathbf{a}}$, $g_1 \wedge \mathbf{a} + g_2 \wedge \mathbf{a} + \cdots + g_n \wedge \mathbf{a}$, and therefore

$$g_r \wedge \overline{\mathbf{a}} = \overline{\mathbf{a}}.$$

Thus each element g_r of the group fixes $\overline{\mathbf{a}}$, as required. ∎

If you have not met the concept of *centroid* in previous work, do not worry. All that matters is that $\overline{\mathbf{a}}$, as here defined, is fixed by all the elements of G (as the proof of the theorem shows).

You saw this proved for $\mathbb{R}^2$ in Theorem 3.2 of *Unit IB1*. The proof given there can easily be adapted to $\mathbb{R}^m$. We supply an alternative, more geometric, proof at the beginning of Section 2 of this unit.

Property (c) of a group action.

In investigating finite groups of isometries, it is convenient to choose the origin of our coordinate system to be fixed by all the elements of the group. Theorem 1.1 allows us to do this, and so *throughout the remainder of this unit we always assume that the origin is fixed by G — that is, G is a group of orthogonal linear transformations.*

We are now in a position to verify that the finite groups of plane isometries are just the cyclic and dihedral groups.

> **Theorem 1.2 Cyclic and dihedral groups**
>
> Let G be a finite group of isometries of $\mathbb{R}^2$. Then, by a suitable choice of coordinate system, we have
>
> *either* G consists only of rotations; that is, $G = C_n = \langle r[2\pi/n] \rangle$
>
> *or* G contains an indirect isometry and $G = D_n = \langle r[2\pi/n], q[0] \rangle$.

Changing the choice of coordinate system is equivalent to *conjugating* by the isometry that maps one origin and axis set to the other. Therefore, the theorem could be rephrased as 'Any finite group G of isometries of $\mathbb{R}^2$ is conjugate to C_n or D_n for some n'.

Proof

Choose the origin O to be a fixed point.

Suppose that G contains no indirect isometries. Then G consists entirely of rotations about O, as these are the only direct isometries that fix O. Then, letting $n = |G|$, we have $G = C_n = \langle r[2\pi/n] \rangle$, from Exercise 1.1.

Suppose now that G does contain an indirect isometry, q. Then q cannot be a glide reflection, as glide reflections have no fixed points. Thus we may choose the x-axis such that $q = q[0]$.

Let H be the subgroup of G consisting of direct isometries; by Exercise 1.1, we have

$$H = C_n = \langle r[2\pi/n] \rangle$$

for some n. Moreover, if s is *any* reflection in G, then $q^{-1}s$, being direct, belongs to H, and so

$$s = q(q^{-1}s) \in \langle r[2\pi/n], q[0] \rangle.$$

Thus,

$$G = D_n = \langle r[2\pi/n], q[0] \rangle,$$

as required. ∎

Exercise 1.2

In the Appendix to *Unit IB4* we described the dihedral group of order $2n$ as follows.

> Each D_n is generated by a rotation r through $2\pi/n$ (so that $r^n = e$) and a reflection s in an axis of symmetry (so that $s^2 = e$).
> In general we have
> $$D_n = \langle r, s : r^n = e, \ s^2 = e, \ sr = r^{n-1}s \rangle.$$

Show that the relation $sr = r^{n-1}s$ may be written in the form $srs^{-1} = r^{-1}$.

We shall find this way of writing the relation more convenient when r and s are written as products of cycles in S_n, because we then have a simple rule for writing down the conjugate srs^{-1}.

1.2 The groups $O(2)$ and $SO(2)$

The set of *all* orthogonal transformations of $\mathbb{R}^2$ is, of course, a group. You met it in passing, in Exercise 1.5 of *Unit IB3*; it is the symmetry group of the 'plane figure' consisting simply of the origin of $\mathbb{R}^2$.

This group is usually written as $O(2)$, and is called the **orthogonal group** in $\mathbb{R}^2$.

The O stands for *orthogonal*, and the 2 refers to the two dimensions of the plane.

The corresponding group of matrices consists of all matrices of the form

$$\begin{bmatrix} \cos\theta & -\sin\theta \\ \sin\theta & \cos\theta \end{bmatrix},$$

which is the matrix of the rotation $r[\theta]$, or of the form

$$\begin{bmatrix} \cos 2\theta & \sin 2\theta \\ \sin 2\theta & -\cos 2\theta \end{bmatrix},$$

which is the matrix of $q[\theta]$.

See Subsection 5.1 of *Unit IB1*.

The notation $O(2)$ is also frequently used to denote this group of matrices.

Now it turns out that $O(2)$ splits into two equal-sized pieces in a way which has an interesting geometrical significance! Remember that a reflection cannot be performed as a rigid motion within $\mathbb{R}^2$. To reflect a thin cardboard shape you must lift it *out* of the plane, turn it over and then replace it. As we have seen earlier in the course, this is sometimes referred to by saying that reflections are *indirect* isometries, whereas rotations are *direct* isometries.

This rather vague geometric notion has an exact algebraic description. Any matrix in $O(2)$ has determinant ± 1, and rotations are distinguished from reflections by having the positive sign. Now the set $\{1, -1\}$ is itself a group under multiplication — it is isomorphic to the cyclic group of order 2, generated by the element -1. So we can consider the function det which maps any orthogonal matrix to its determinant as a function from the group $O(2)$ to the cyclic group C_2:

$$\det : O(2) \to C_2.$$

If G is a subgroup of $O(2)$, then det also defines (simply by restriction to G) a function from G to C_2. In fact det is no mere function — it is a group homomorphism! We shall show this presently.

The *kernel* of det — the subgroup of elements of G which map to the identity, $+1$, of C_2 — is the subgroup of *direct* isometries (rotations) in G.

This discussion can be formalized as follows.

Theorem 1.3

Let G be a subgroup of $O(2)$. The function $\det : G \to C_2$ is a homomorphism. The kernel of det is either the whole of G or a (normal) subgroup of G of index 2.

Proof

For square matrices $\mathbf{A}$, $\mathbf{B}$, we always have $\det \mathbf{AB} = \det \mathbf{A} \det \mathbf{B}$, so det is indeed a homomorphism from G to C_2.

The second part is an immediate consequence of a well-known theorem of group theory, but in this case we can also see the result geometrically.

First Isomorphism Theorem (Theorem 4.5 in *Unit IB4*).

First note that the rotations in G certainly form a subgroup G^+. If $\det \mathbf{A} = +1$ and $\det \mathbf{B} = +1$, then certainly $\det \mathbf{AB} = +1$, so G^+ is closed. Also, the determinant of the identity matrix is 1, so the identity matrix belongs to G^+. Finally, for any $\mathbf{A}$ in G, $\det \mathbf{A}^{-1}$ has the same sign as $\det \mathbf{A}$, so G^+ contains the inverse of each of its elements.

The kernel of det is the subgroup G^+, since each element of G^+ maps to $+1$, the identity in C_2, and any reflections belonging to G map to -1. Thus, if G contains no reflections, then the kernel of det is the whole of G.

Suppose that there is at least one reflection s in G. To show that G^+ has index 2 we show that there are just two left cosets of G^+ in G, so that

$$G = G^+ \cup sG^+.$$

If G consists only of $\{e, s\}$ this is obvious, since $G^+ = \{e\}$. So suppose that s' is another reflection in G. We shall show that $s' = sr$, where $r \in G^+$. All we need to observe is that a reflection has order 2, and so $s^{-1} = s$. Thus the product

$$s^{-1}s' = ss'$$

is a rotation $r \in G^+$, since $\det ss' = (-1)(-1) = +1$. So $s' = sr$, as required.

Thus G splits into two pieces: the subgroup G^+ consisting of rotations, and the set G^- ($= sG^+$) of reflections. ∎

Exercise 1.3

Why is it impossible for G^- to be a group?

If G happens to be the whole of $O(2)$, the subgroup $O^+(2)$ of direct isometries is the group of *all* rotations of $\mathbb{R}^2$ which fix the origin, and this is usually called the **special orthogonal group** in $\mathbb{R}^2$ — written $SO(2)$ for short.

We can now interpret Theorem 1.3 as a result about finite subgroups of $O(2)$ and $SO(2)$.

> **Theorem 1.4**
>
> (a) Every finite subgroup of $SO(2)$ is equal to C_n for some n.
>
> (b) Every finite subgroup of $O(2)$ is *either* equal to C_n for some n or conjugate to D_n for some n.

Proof

Part (a) follows directly from Exercise 1.1, as does part (b) in the case of a finite subgroup of $O(2)$ which is contained in $SO(2)$.

Suppose, then, that G is a finite subgroup of $O(2)$ which contains some indirect orthogonal transformation $q[\theta]$. Consider the conjugate $\widetilde{G}$ of G by $r[-\theta]$:

$$\widetilde{G} = r[-\theta]\, G\, (r[-\theta])^{-1}$$
$$= r[-\theta]\, G\, r[\theta]. \qquad \square$$

Exercise 1.4

Show that $\widetilde{G}$ contains $q[0]$.

Proof of Theorem 1.4 continued

Let $\widetilde{G}^+ = G^+$ be the subgroup of G consisting of elements of $SO(2)$. By Theorem 1.2, $\widetilde{G}^+ = C_n$ for some n.

Now, arguing as in the proof of Theorem 1.2,

$$\widetilde{G} = \langle r[2\pi/n], q[0] \rangle$$
$$= D_n.$$

This completes the proof. ∎

Often it is interesting to try to find a geometric figure which has a given group as its group of symmetries. For the group D_n of Theorem 1.4, provided $n > 2$, we can choose such a figure to be the regular n-gon. Notice that the possible axes of reflection are of *three* different types (see Figure 1.1).

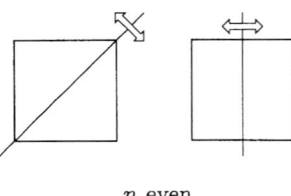

n even $\qquad\qquad$ n odd

Figure 1.1

If n is even, such an axis either passes through two opposite vertices, or through the midpoints of opposite edges. If n is odd, it passes through one vertex and the midpoint of the opposite edge.

In this context, the cyclic group C_n is the *subgroup* of rotations of D_n. However, we can create figures which have C_n as their *full* symmetry group by removing any reflective symmetry. For example, by adding some markings to a square we obtain a figure whose full symmetry group is C_4, and doing the same to an equilateral triangle we get a figure with full symmetry group C_3 (see Figure 1.2).

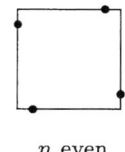 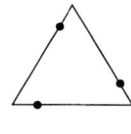

n even n odd

Figure 1.2

Where necessary we shall refer to such figures as *marked n-gons*.

There is one final remark to make before we step into the third dimension! It may not appear significant now, but later you will see its importance.

Definition 1.1 Central inversion

Central inversion of $\mathbb{R}^m$ is the isometry σ_O defined by

$\sigma_O : \mathbf{x} \mapsto -\mathbf{x}.$

Strictly speaking, σ_O is a different isometry for each m, but as the appropriate value of m will always be clear from the context, this should not cause any problems.

Informally, this means that any point is mapped 'straight through the origin to a point the same distance the other side'.

Exercise 1.5

Show that in $\mathbb{R}^2$ central inversion is the same as a rotation through π; that is, show that in $\mathbb{R}^2$

$\sigma_O = r[\pi].$

2 ISOMETRIES IN THREE DIMENSIONS

We now progress from $\mathbb{R}^2$ into $\mathbb{R}^3$; the study of finite symmetry groups of $\mathbb{R}^3$ will occupy the remainder of this unit.

2.1 Isometries in $\mathbb{R}^3$

Many of the ideas we have already met are directly applicable. We commence with a theorem which fully describes three-dimensional isometries.

Theorem 2.1

Any isometry of $\mathbb{R}^3$ consists of a linear transformation represented by an orthogonal matrix (which fixes the origin), followed by a translation.

The proof for $\mathbb{R}^2$ given in Section 3 of *Unit IB1* can in fact be generalized to $\mathbb{R}^m$ very easily; but this proof is not geometrically very revealing. We shall now give a geometric proof of Theorem 2.1, explicitly for $\mathbb{R}^3$.

Proof

Let the image of the origin O under the isometry ϕ be $\mathbf{a}$. Consider the isometry ψ given by

$$\psi(\mathbf{x}) = \phi(\mathbf{x}) - \mathbf{a}.$$

It is clear that this fixes the origin, i.e. $\psi(\mathbf{0}) = \mathbf{0}$. Since an isometry preserves distance between points, any point with distance 1 from the origin has an image under ψ which is also distance 1 from $\mathbf{0}$. Therefore the unit sphere is mapped to itself by ψ.

The unit sphere is the spherical surface whose points are at unit distance from the origin in $\mathbb{R}^3$.

Now let $\psi(0,0,1) = \mathbf{b}$ (see Figure 2.1).

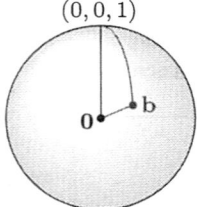

Figure 2.1

Choose a rotation R of the sphere which sends $\mathbf{b}$ to $(0,0,1)$. Then

$$\lambda = R\psi$$

is an isometry which fixes the origin and $(0,0,1)$.

Now, just as for the two-dimensional case, an isometry of $\mathbb{R}^3$ which fixes the origin preserves the *dot product*. In other words, for all $\mathbf{a}, \mathbf{b} \in \mathbb{R}^3$,

$$\lambda(\mathbf{a}) \cdot \lambda(\mathbf{b}) = \mathbf{a} \cdot \mathbf{b}.$$

We ask you to verify this in Exercise 2.1.

Now any vector $\mathbf{a}$ lying in the xy-plane of $\mathbb{R}^3$ has the property that $\mathbf{a} \cdot (0,0,1) = 0$. Therefore, since λ fixes $(0,0,1)$,

$$\lambda(\mathbf{a}) \cdot (0,0,1) = 0.$$

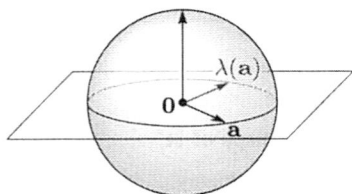

Figure 2.2

Thus, in addition to fixing the origin and $(0,0,1)$, λ maps the xy-plane to itself. Thus we know that the effect of λ on the xy-plane is either a rotation or a reflection. Furthermore, a general vector $\mathbf{v} = (x,y,z) \in \mathbb{R}^3$ has $\mathbf{v} \cdot (0,0,1) = z$. Thus $\lambda(\mathbf{v}) \cdot (0,0,1) = z$, and λ is 'height-preserving'!

The effect of λ on the xy-plane is of a two-dimensional isometry that fixes the origin in the plane; we saw in Unit IB1 that such a two-dimensional isometry must be of the form $r[\theta]$ or $q[\theta]$.

The upshot of this discussion is that the effect of λ on a general point (x,y,z) is to preserve z and transform x and y by either a rotation about the z-axis or a reflection in a plane perpendicular to the xy-plane and containing the z-axis.

These two possibilities are represented by the matrices

$$\begin{bmatrix} \cos\theta & -\sin\theta & 0 \\ \sin\theta & \cos\theta & 0 \\ 0 & 0 & 1 \end{bmatrix} \quad \text{and} \quad \begin{bmatrix} \cos 2\theta & \sin 2\theta & 0 \\ \sin 2\theta & -\cos 2\theta & 0 \\ 0 & 0 & 1 \end{bmatrix},$$

which represent anticlockwise rotation through θ about the z-axis and reflection in the vertical plane $x\sin\theta - y\cos\theta = 0$, respectively. Notice that both of these are orthogonal.

> You are asked to check this in Exercise 2.2.

As before, we recover the original isometry ϕ by undoing the steps, so that

$$\phi = R^{-1}\lambda + \mathbf{a}.$$

To complete the proof, all we need to do is show that $R^{-1}\lambda$ is represented by an orthogonal matrix. Since R is a rotation, the image of the three standard basis vectors $(1,0,0), (0,1,0)$ and $(0,0,1)$ of $\mathbb{R}^3$ under R is a set of three mutually perpendicular unit vectors. These are the columns of the matrix representing R, which is therefore orthogonal. Therefore the matrix representing R^{-1} is also orthogonal, and hence so is the matrix for $R^{-1}\lambda$. This completes the proof. ∎

Exercise 2.1

Verify that if ϕ is an isometry which fixes the origin then, for all $\mathbf{a}, \mathbf{b} \in \mathbb{R}^3$, $\phi(\mathbf{a}) \cdot \phi(\mathbf{b}) = \mathbf{a} \cdot \mathbf{b}$.

Exercise 2.2

Check that the transformation of $\mathbb{R}^3$ determined by the matrix

$$\begin{bmatrix} \cos 2\theta & \sin 2\theta & 0 \\ \sin 2\theta & -\cos 2\theta & 0 \\ 0 & 0 & 1 \end{bmatrix}$$

is a reflection in the vertical plane $x\sin\theta - y\cos\theta = 0$.

The group of all 3×3 orthogonal matrices corresponds precisely to the set of isometries of $\mathbb{R}^3$ which fix the origin. It is written $O(3)$, and is called the **orthogonal group** in $\mathbb{R}^3$; we shall be interested in finite subgroups of this group.

In order to facilitate our search for these subgroups, we use some of the ideas from Section 1 — in particular, the Fixed Point Theorem (Theorem 1.1).

2.2 Finite groups of isometries in $\mathbb{R}^3$

From now on, for any finite group G of isometries of $\mathbb{R}^3$, we take the fixed point of G as the origin. Thus, by Theorem 2.1, every element of G can be represented by a 3×3 orthogonal matrix, and for the remainder of this unit the word 'isometry' may be interpreted as 'element of $O(3)$'. Thus we no longer admit translations.

Just as in the two-dimensional case, the group $O(3)$ consists of two parts — the *direct* and the *indirect* isometries. To see the distinction, consider what happens to the unit vectors $\mathbf{i}, \mathbf{j}$ and $\mathbf{k}$ under the different types of isometries.

Let g be an element of $O(3)$, and denote the images $g(\mathbf{i}), g(\mathbf{j})$ and $g(\mathbf{k})$ by $\mathbf{i}'$, $\mathbf{j}'$ and $\mathbf{k}'$ respectively. Two possible image sets are shown in Figure 2.3.

> Throughout the remainder of this unit the vectors $(1,0,0), (0,1,0)$ and $(0,0,1)$ will be denoted by $\mathbf{i}, \mathbf{j}$ and $\mathbf{k}$, respectively.

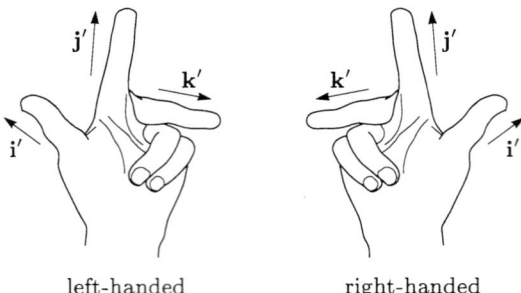

left-handed   right-handed

Figure 2.3

We shall say that $\{\mathbf{i}',\mathbf{j}',\mathbf{k}'\}$ forms a *left-handed* or a *right-handed* **orthonormal basis** of $\mathbb{R}^3$ according to which hand can comfortably be held so that the thumb, forefinger and middle finger point in the directions of $\mathbf{i}'$, $\mathbf{j}'$ and $\mathbf{k}'$ respectively (see Figure 2.3). Since $\{\mathbf{i},\mathbf{j},\mathbf{k}\}$ forms a right-handed basis, it follows that g is direct if $\{\mathbf{i}',\mathbf{j}',\mathbf{k}'\}$ is right-handed and indirect if $\{\mathbf{i}',\mathbf{j}',\mathbf{k}'\}$ is left-handed. In the former case, the determinant of the matrix is $+1$, whereas in the latter case it is -1.

Intuitively, the only rigid motions which you can actually perform on an object in space correspond to direct isometries. The indirect isometries require space to 'turn inside out', rather like turning a right-handed glove into a left-handed one.

The distinction between the cases where det $= +1$ and det $= -1$ distinguishes direct from indirect isometries in any number of dimensions.

The direct isometries of $\mathbb{R}^3$ that fix O (corresponding to orthogonal matrices with determinant $+1$) form a subgroup of $O(3)$, called the **special orthogonal group** in $\mathbb{R}^3$ and denoted by $SO(3)$. It turns out (as you will see in a corollary to Theorem 2.3) that, as in the two-dimensional case, each element of $SO(3)$ is a rotation.

The classification of *indirect* isometries in $O(3)$ is more complicated. *Some* of them are (as we might expect) reflections in planes through the origin (see Figure 2.4).

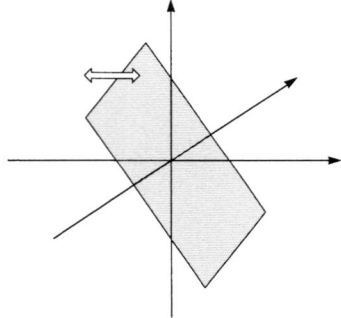

Figure 2.4

It turns out that by no means all indirect isometries are reflections; but any indirect isometry is *either* a reflection *or* the composite of three reflections, while any direct isometry is the composite of two reflections. We shall now formally state and prove this.

Recall that we are considering only those that fix the origin.

Theorem 2.2

Any isometry ϕ in $O(3)$ is the composite of at most three reflections, each in a plane through the origin. Furthermore, ϕ is the composite of just two such reflections if and only if ϕ is direct.

Proof

If $\phi = e$, then $\phi = s \circ s$ for *any* reflection s. Thus, we may assume that $\phi \neq e$. Therefore there is a unit vector $\mathbf{p}$ such that $\phi(\mathbf{p}) \neq \mathbf{p}$. Denote $\phi(\mathbf{p})$ by $\mathbf{p}'$, and let P be the plane through the midpoint of the line segment $\mathbf{pp}'$ and perpendicular to this line segment. Since $\mathbf{p}$ and $\mathbf{p}'$ are both unit vectors, P passes through the origin (see Figure 2.5).

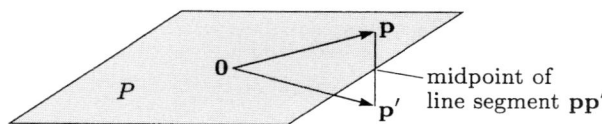

Figure 2.5

Let s_1 be reflection in the plane P. Then $s_1(\mathbf{p}') = \mathbf{p}$, and so the composite $s_1\phi$ fixes both $\mathbf{0}$ and $\mathbf{p}$.

Now let S be the plane through $\mathbf{0}$ containing $\mathbf{p}$ and $\mathbf{p}'$, and let $\mathbf{q}$ be a unit vector in S, perpendicular to $\mathbf{p}$. Let $\mathbf{q}' = (s_1\phi)(\mathbf{q})$, and let Q be the plane passing through $\mathbf{0}$, $\mathbf{p}$ and the midpoint of the line segment $\mathbf{qq}'$, see Figure 2.6.

Even if $\mathbf{q}' = \mathbf{q}$, Q is well-defined; it is the plane through $\mathbf{0}$, $\mathbf{p}$ and $\mathbf{q}$ in this case.

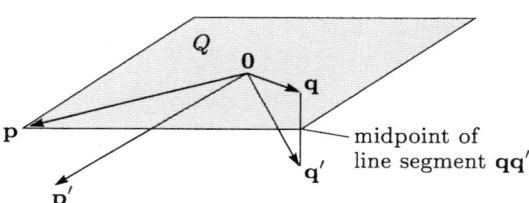

Figure 2.6

Since $s_1\phi$ fixes $\mathbf{p}$, the vectors $\mathbf{q}$ and $\mathbf{q}'$ must both be perpendicular to $\mathbf{p}$, and hence Q is perpendicular to the line segment $\mathbf{qq}'$ (unless $\mathbf{q} = \mathbf{q}'$). Thus, whether or not $\mathbf{q} = \mathbf{q}'$, the reflection s_2 in the plane Q must map $\mathbf{q}'$ to $\mathbf{q}$.

Thus the composite isometry $s_2 s_1 \phi$ fixes $\mathbf{0}$, $\mathbf{p}$ and $\mathbf{q}$. Let $\mathbf{r}$ be a vector perpendicular to both $\mathbf{p}$ and $\mathbf{q}$; then $s_2 s_1 \phi$ can map $\mathbf{r}$ only to $\mathbf{r}$ or to $-\mathbf{r}$.

This is because $(s_2 s_1 \phi)(\mathbf{r})$ must be the same length as $\mathbf{r}$, and must be perpendicular to $\mathbf{p}$ and $\mathbf{q}$.

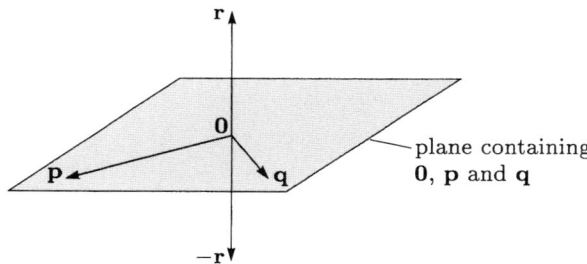

Figure 2.7

In the former case, $s_2 s_1 \phi$ is the identity isometry, while in the latter case it is the reflection s_3 in the plane containing $\mathbf{0}$, $\mathbf{p}$ and $\mathbf{q}$. Thus:

either $s_2 s_1 \phi = e$, in which case $\phi = s_1^{-1} s_2^{-1} = s_1 s_2$;

or $s_2 s_1 \phi = s_3$, in which case $\phi = s_1^{-1} s_2^{-1} s_3 = s_1 s_2 s_3$.

This shows that, at worst, ϕ is the composite of three reflections, as claimed. Clearly ϕ is direct if and only if it is the composite of an even number of reflections (i.e. two reflections). ∎

From this result we can easily show that any non-trivial direct isometry in $SO(3)$ must be a rotation.

As with isometries in $\mathbb{R}^2$, we regard the identity as a 'trivial' rotation.

> ***Theorem 2.3***
>
> The composite of two reflections in distinct planes through the origin is a rotation about an axis through the origin consisting of the intersection of the planes.

Proof

Figure 2.8 shows two distinct planes P_1 and P_2 through the origin.

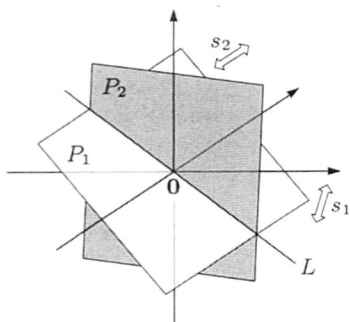

Figure 2.8

They meet along a line L through $\mathbf{0}$. Let s_1 and s_2 be the corresponding reflections. Since the points of L are fixed by s_1 and s_2, they are fixed by the composite $s_2 s_1$. Then the plane P through $\mathbf{0}$ perpendicular to L is mapped to itself by $s_2 s_1$. Thus the restriction of $s_2 s_1$ to P is a two-dimensional isometry fixing the origin. In fact, it is the composite of the two-dimensional reflections in the lines in P where P_1 and P_2 intersect P. Therefore $s_2 s_1$ is a rotation of the plane P. Since the points of L are mapped to themselves, $s_2 s_1$ can only be a rotation with axis L. ∎

Combining this result with Theorem 2.2 gives the following corollary.

> ***Corollary 2.1***
>
> Any direct isometry fixing the origin is a rotation.

Thus far, the situation is similar to the two-dimensional case, although, as we shall soon see, a finite rotation group in $\mathbb{R}^3$ need *not* be cyclic.

The changes become more marked when we examine *indirect* isometries. The next theorem gives a useful characterization of these; but before studying this, it will be useful for you to derive for yourself some important properties of the particular indirect isometry known as central inversion.

Exercise 2.3 ─────────

Prove that, in $\mathbb{R}^3$, σ_O cannot be represented as a single reflection.

Exercise 2.4 ─────────

Show that, given *any* plane P through the origin, σ_O is the composite of reflection in P with rotation through π about the axis through the origin perpendicular to P.

Exercise 2.5

By representing σ_O as a matrix, show that σ_O commutes with every other element of $O(3)$.

We can now characterize *any* indirect isometry in $O(3)$ as the composite of a reflection in a plane with a rotation about the perpendicular to that plane.

> **Theorem 2.4**
>
> An indirect isometry fixing the origin can be expressed as the composite of reflection in a plane P through $\mathbf{0}$ with a rotation about the axis through $\mathbf{0}$ perpendicular to P.

Proof

Let s be any indirect isometry fixing $\mathbf{0}$. Consider the composite $s\sigma_O$; let us denote this composite by ρ.

Since σ_O and s are each indirect, ρ is direct (and fixes $\mathbf{0}$). Therefore, by Corollary 2.1, ρ is a rotation.

Let P be the plane through $\mathbf{0}$ perpendicular to the axis of rotation of ρ. By the result of Exercise 2.4, we can write σ_O as

$$\sigma_O = r\, q$$

where q is reflection in P and r is rotation through π about the same axis as that of ρ.

Now, as $\rho = s\,\sigma_O$ and $\sigma_O^2 = e$, we have

$$s = \rho\,\sigma_O$$
$$= \rho\, r\, q,$$

and as r and ρ are rotations about the same axis, their composite is another such rotation. Thus,

$$s = (\rho\, r)\, q$$

is the expression we are looking for. ∎

The final result in this subsection returns to the subject of *groups* of isometries (rather than characterizations of individual isometries). It is the analogue for $\mathbb{R}^3$ of Theorem 1.2.

> **Theorem 2.5**
>
> Let G be a finite subgroup of $O(3)$, and let G^+ denote the subgroup of G consisting of rotations. Then:
>
> *either* G consists only of rotations; that is, $G = G^+$;
>
> *or* G contains an indirect isometry σ, and $G = G^+ \cup \sigma G^+$.

Proof

The crucial fact is that (as we observed in proving Theorem 2.4) the product of two indirect isometries is direct. If G consists only of rotations, there is nothing to prove; so suppose that G contains an indirect isometry, σ.

All the elements of σG^+ belong to G, and so

$$G \supseteq G^+ \cup \sigma G^+.$$

To prove the reverse inclusion, let λ be *any* indirect isometry in G. Then $\sigma^{-1}\lambda$, being direct, belongs to G^+. Thus

$$\lambda = \sigma(\sigma^{-1}\lambda) \in \sigma G^+,$$

and so $G \subseteq G^+ \cup \sigma G^+$. We have established the inclusion both ways, so the theorem is proved. ∎

2.3 Reflections of $\mathbb{R}^2$ can be rotations in $\mathbb{R}^3$

None of our efforts in revising symmetry groups of $\mathbb{R}^2$ have been in vain. In fact every finite symmetry group acting on $\mathbb{R}^2$ also acts on $\mathbb{R}^3$, which gives us a very good start. The reason is simple: $\mathbb{R}^2$ fits naturally 'inside' $\mathbb{R}^3$. One obvious way this can be done is to embed $\mathbb{R}^2$ as the xy-plane of $\mathbb{R}^3$, see Figure 2.9.

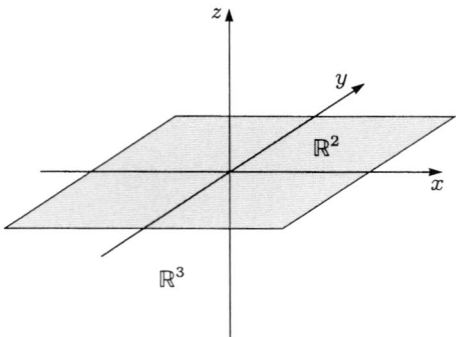

Figure 2.9

Then any symmetry of $\mathbb{R}^2$ can be extended to $\mathbb{R}^3$. The only complication is that increasing the dimension involves some different *geometric* interpretations.

The first important observation is that any *reflection* of a plane figure embedded in $\mathbb{R}^3$ in this way can be achieved simply by a *rotation* of $\mathbb{R}^3$.

To see this, consider the particular example of the symmetry group of an equilateral triangle, shown in Figure 2.10. We have seen that this group is isomorphic to D_3 (which also happens to be isomorphic to S_3). When the plane $\mathbb{R}^2$ is embedded in $\mathbb{R}^3$ as above, we find that the 'reflection' which interchanges vertices 2 and 3 can be realized as a *rotation* of $\mathbb{R}^3$ through π about the axis joining 1 to the midpoint of 23.

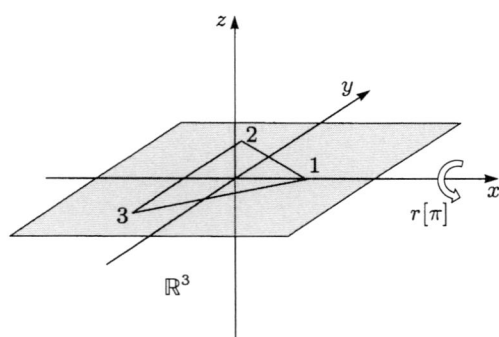

Figure 2.10

Of course, it could also be realized by the $\mathbb{R}^3$ reflection in the zx-plane.

From now onwards we shall need a clear distinction between the direct isometries (rotations) and indirect isometries of $\mathbb{R}^3$. To assist in this we shall henceforth use the following notation for the symmetries of a *solid figure F*, that is, a subset of $\mathbb{R}^3$.

> **Notation** **Symmetries of a solid figure F**
>
> $\Gamma^+(F)$ the group of rotational (direct) symmetries of F
>
> $\Gamma^-(F)$ the set of indirect symmetries of F
>
> $\Gamma(F)$ the full symmetry group (both direct and indirect) of F

This is an extension of the notation which we set up in *Unit IB3*.

Exercise 2.6

Show that if a group G of symmetries of $\mathbb{R}^3$ contains only one indirect symmetry, then it is isomorphic to the dihedral group D_1.

3 DIRECT GROUPS

Throughout this section we shall confine our attention to rotations of $\mathbb{R}^3$. As before, this will be a good base from which we can later investigate general symmetry groups. We are interested in symmetries of solid figures rather than planar polygons, and so we shall abandon regular polygons in favour of some solid figures. The idea is to think of the unit sphere in $\mathbb{R}^3$.

Now, instead of dealing with a regular planar n-gon, we mark n equally spaced points on the equator of the sphere — so dividing the equator into equal arcs. The northern and southern hemispheres can be thought of as *faces* which are joined together along the equator (see Figure 3.1).

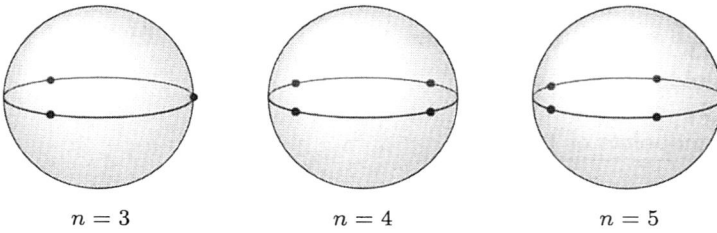

$n = 3$ $n = 4$ $n = 5$

Figure 3.1

This marked sphere is sometimes referred to as a **dihedron**, which is the origin of the term 'dihedral'. Throughout this unit we shall use the symbol DIH to denote dihedron, and the number of equatorial vertices will be written as a subscript, i.e. DIH_n.

It will often be useful to be able to distinguish the two faces, for example to deal with reflection in the xy-plane (a non-trivial symmetry of $\mathbb{R}^3$ which is the identity on the copy of $\mathbb{R}^2$ consisting of the xy-plane). Therefore we shall usually mark the north and south poles, N and S, on the sphere. DIH_3 so marked is shown in Figure 3.2.

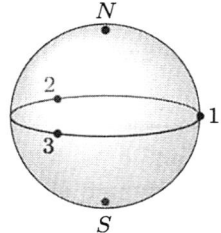

Figure 3.2

Labelling the poles in this way is equivalent to labelling the two faces of the dihedron.

Now consider an $\mathbb{R}^2$ symmetry of the regular n-gon. The n-gon sits inside the equatorial plane of the dihedron, and the symmetry defines a unique *rotational* symmetry of the dihedron DIH_n in $\mathbb{R}^3$. We have seen that even an $\mathbb{R}^2$ reflection of the n-gon can be achieved by a *rotation* of $\mathbb{R}^3$. The converse is also true: any rotational symmetry of the dihedron defines a unique $\mathbb{R}^2$ symmetry of the equatorial n-gon. This follows from the fact that either symmetry is completely determined by its effect on the vertices.

Example 3.1

We give a detailed description of $\Gamma^+(\text{DIH}_4)$, the rotation group of DIH_4 (see Figure 3.3).

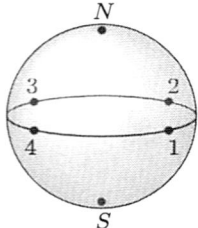

Figure 3.3

We can write the rotations of $\Gamma^+(\text{DIH}_4)$ explicitly as cycles, as in Table 3.1.

Element	Name	Angle	Axis
e	e	0	any
(1234)	r	$\pi/2$	NS
$(13)(24)$	r^2	π	NS
(1432)	r^{-1}	$-\pi/2$	NS
$(NS)(24)$	ρ	π	13
$(NS)(14)(23)$	ρr	π	midpoints of 14, 23
$(NS)(13)$	ρr^2	π	24
$(NS)(12)(34)$	ρr^{-1}	π	midpoints of 12, 34

Table 3.1

We see that there are two generators, r and ρ, and that

$$r^4 = \rho^2 = e.$$

Moreover,

$$\rho r \rho^{-1} = (NS)(24)(1234)(NS)(24) = (1432) = r^{-1},$$

and so $\Gamma^+(\text{DIH}_4)$ is isomorphic to the dihedral group D_4. ♦

Here, we have written the relation $\rho r = r^{n-1}\rho$ of a dihedral group with generator r of order n and generator ρ of order 2 in the form

$$\rho r \rho^{-1} = r^{-1}$$

(see Exercise 1.2).

A similar argument works for a general dihedron DIH_n, and thus we have an important and memorable start towards our search for finite symmetry groups of $\mathbb{R}^3$, given in the following theorem.

Theorem 3.1

The group of rotational symmetries of a dihedron is dihedral:

$$\Gamma^+(\text{DIH}_n) \cong D_n.$$

Exercise 3.1

Which simple restriction on the type of rotation gives the cyclic subgroup C_n of D_n?

To form the dihedron DIH_2 in $\mathbb{R}^3$ we proceed as before, but use only two vertices. The resulting figure, Figure 3.4, has two vertices, two edges and two faces. As before, we label the faces N and S but, in this case only, it is convenient to label the edges as well: we use the labels E and W.

The edges are the equatorial arcs joining the equatorial vertices.

The labels E and W are not needed to distinguish between the elements of $\Gamma^+(\text{DIH}_2)$, but they *are* needed in order to distinguish (for example) between reflection in the equatorial plane, represented by (NS), and rotation about 12, represented by $(NS)(EW)$.

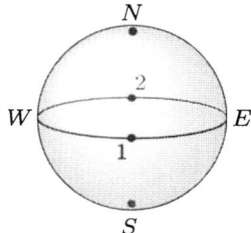

Figure 3.4

The four rotations of DIH_2 can now be described as:

e,
$(NS)(EW)$ rotation through π about 12,
$(12)(EW)$ rotation through π about NS,
$(NS)(12)$ rotation through π about EW.

The regular solids

You are advised to watch the video programme VC4A, *The regular solids*, at this point in your study of the unit.

Please turn now to the video notes for this programme, do the pre-video work indicated there, watch the video programme, and then do the post-video work before continuing your study of this unit.

To summarize the work in the video, we can now write down the direct symmetry groups of the regular tetrahedron, cube and regular dodecahedron as follows:

$$\Gamma^+(\text{TET}) \cong A_4, \quad \Gamma^+(\text{CUBE}) \cong S_4, \quad \Gamma^+(\text{DODECA}) \cong A_5.$$

4 FINITE SUBGROUPS OF $O(3)$

4.1 The coset decomposition revisited

We now wish to consider finite subgroups of $O(3)$. These will turn out to include the full symmetry groups of the various objects (dihedron, tetrahedron, cube, octahedron, dodecahedron and icosahedron) we have studied, plus a few surprises! Of course, the new ingredients we are after involve *indirect* symmetries like reflections. In this section we shall concentrate on the regular solids, including the dihedra. Our search is aided by Theorem 2.5, which tells us that such a symmetry group G (assuming that it is finite) *either* consists entirely of rotations (i.e. $G = G^+$) *or* takes the form $G^+ \cup \sigma G^+$ where σ is any indirect symmetry in G.

If we can choose σ so that it is of order 2 and commutes with every element of G^+, then we obtain the following useful characterization.

Theorem 4.1

Let G be a finite subgroup of $O(3)$, containing an indirect isometry σ of order 2 which commutes with every element of G^+. Then

$G \cong G^+ \times C_2$.

Proof

We use Theorem 1.1 of *Unit GR2*, which states that a function

$$\phi : H_1 \times H_2 \to G$$
$$(h_1, h_2) \mapsto h_1 h_2$$

is an isomorphism if and only if the following three conditions hold:

(a) $G = H_1 H_2$;
(b) $H_1 \cap H_2 = \{e\}$;
(c) H_1 and H_2 are normal subgroups of G.

In our case, we take $H_1 = G^+$ and $H_2 = \{e, \sigma\}$, defining ϕ by

$$\phi : G^+ \times \{e, \sigma\} \to G$$
$$(h, e) \mapsto h$$
$$(h, \sigma) \mapsto \sigma h.$$

It remains to verify the three conditions.

Condition (a) follows directly from Theorem 2.5, while Condition (b) follows from the fact that σ is an indirect isometry and is therefore disjoint from G^+. To verify Condition (c), we need to show that G^+ and $\{e, \sigma\}$ are normal subgroups of G.

Now for any $g \in G^+$ and any $h \in G$, the isometry hgh^{-1} is clearly a direct isometry, so $hG^+h^{-1} \subseteq G^+$. This shows that G^+ is normal. The fact that σ commutes with every element of G^+ (and hence with every element of G) shows that $\{e, \sigma\}$ is normal.

It follows that ϕ is an isomorphism; that is, Note that $\{e, \sigma\} \cong C_2$.

$$G \cong G^+ \times C_2.$$ ∎

We shall now deal the dihedra and the regular solids, in turn.

4.2 The dihedron with n equatorial vertices

We know from Subsection 4.1 that $\Gamma(\mathrm{DIH}_n)$ can be decomposed as $\Gamma^+(\mathrm{DIH}_n) \cup \sigma\Gamma^+(\mathrm{DIH}_n)$, where σ is any indirect symmetry. We also know that, if we can choose σ to be a reflection that commutes with all of $\Gamma^+(\mathrm{DIH}_n)$, then we can apply Theorem 4.1.

There *is* such a choice for σ, namely reflection in the xy-plane. The effect of σ on the vertices of DIH_n is merely to exchange N and S; and since every element of $\Gamma^+(\mathrm{DIH}_n)$ permutes the equatorial vertices among themselves and either fixes or exchanges N and S, it follows that σ does indeed commute with all of $\Gamma^+(\mathrm{DIH}_n)$. This, in conjunction with Theorem 3.1, gives the following theorem.

Theorem 4.2

For any $n \geq 1$, $\Gamma(\mathrm{DIH}_n)$ is isomorphic to $D_n \times C_2$.

We shall illustrate this theorem with the particular case $n = 3$, which is shown in Figure 4.1.

Example 4.1

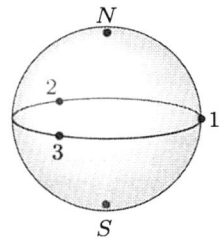

Figure 4.1

The elements of $\Gamma^+(\mathrm{DIH}_3)$ are easily listed as cycles, shown in Table 4.1.

Element	Name	Angle	Axis
e	e	0	any
(123)	r	$2\pi/3$	NS
(132)	r^{-1}	$-2\pi/3$	NS
$(NS)(23)$	ρ	π	1 to midpoint of 23
$(NS)(13)$	ρr	π	2 to midpoint of 13
$(NS)(12)$	ρr^{-1}	π	3 to midpoint of 12

Table 4.1

We know that this group is isomorphic to D_3, and that $\qquad$ Theorem 3.1.
$$r^3 = \rho^2 = e.$$

Now, by Theorem 2.5, composing these six elements with any indirect symmetry σ in $\Gamma^-(\mathrm{DIH}_3)$ will give the remaining six elements of $\Gamma(\mathrm{DIH}_3)$.

In order to apply Theorem 4.1, we choose σ to be a reflection in the xy-plane, which in cycle form is just (NS). A complete list of all the elements of $\Gamma(\mathrm{DIH}_3)$ is given in Table 4.2.

Element	Name	Element	Name
e	e	(NS)	σ
(123)	r	$(NS)(123)$	σr
(132)	r^{-1}	$(NS)(132)$	σr^{-1}
$(NS)(23)$	ρ	(23)	$\sigma\rho$
$(NS)(13)$	ρr	(13)	$\sigma\rho r$
$(NS)(12)$	ρr^{-1}	(12)	$\sigma\rho r^{-1}$

Table 4.2

We can actually express DIH_3 as $D_3 \times C_2$, and thus verify Theorem 4.1, in two different ways! The subgroup $\Gamma^+(\mathrm{DIH}_3)$, namely

$$\{e, (123), (132), (NS)(23), (NS)(13), (NS)(12)\},$$

is isomorphic to D_3 and commutes with $\{e, (NS)\}$; but so is the subgroup

$$\{e, (123), (132), (23), (13), (12)\}. \qquad \blacklozenge$$

4.3 The even and odd cases

Theorem 4.2 works for all values of n; but there are nevertheless subtle differences between the cases n even and n odd, which we shall now investigate.

Case 1 n is even

If n is even, then central inversion σ_O is a symmetry of DIH_n. Now we know from Exercise 2.5 that σ_O commutes with every other element of $O(3)$, and in particular with every other element of $\Gamma(\text{DIH}_n)$. Therefore, we do not need to choose σ to be reflection in the xy-plane in order to apply Theorem 4.1; in the case where n is even, we may alternatively choose $\sigma = \sigma_O$. We shall now work through the case $n = 4$ in detail, with this choice of σ (see Figure 4.2).

Example 4.2

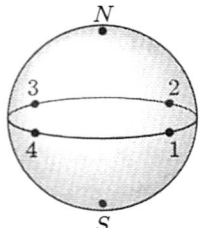

Figure 4.2

We have already seen the rotations of DIH_4 expressed explicitly as cycles, as shown in Table 4.3.

See Example 3.1.

Element	Name	Angle	Axis
e	e	0	any
(1234)	r	$\pi/2$	NS
$(13)(24)$	r^2	π	NS
(1432)	r^{-1}	$-\pi/2$	NS
$(NS)(24)$	ρ	π	13
$(NS)(14)(23)$	ρr	π	midpoints of 14, 23
$(NS)(13)$	ρr^2	π	24
$(NS)(12)(34)$	ρr^{-1}	π	midpoints of 12, 34

Table 4.3

We saw in Example 3.1 that this is isomorphic to the dihedral group D_4. Suitable generators are

$$r = (1234) \text{ of order } 4 \quad \text{and} \quad \rho = (NS)(24) \text{ of order } 2.$$

In cycle notation, the central inversion σ_O is $(NS)(13)(24)$. Thus we can compute the elements of $\sigma_O \Gamma^+(\text{DIH}_4)$.   ◊

Exercise 4.1

Draw up a list of the elements of $\sigma_O \Gamma^+(\text{DIH}_4)$ in cycle form.

Example 4.2 continued

It is clear from Figure 4.2 that each of the symmetries in $\sigma_O \Gamma^+(\text{DIH}_4)$ is indeed an indirect symmetry. Thus for $\Gamma(\text{DIH}_4)$ we have a complete listing of the sixteen elements and, as this expresses $\Gamma(\text{DIH}_4)$ as

$$\Gamma^+(\text{DIH}_4) \cup \sigma_O \Gamma^+(\text{DIH}_4),$$

it follows immediately from Theorem 4.1 that

$$\Gamma(\text{DIH}_4) \cong \Gamma^+(\text{DIH}_4) \times C_2.$$

But $\Gamma^+(\text{DIH}_4) \cong D_4$, so we have

$$\Gamma(\text{DIH}_4) \cong D_4 \times C_2.$$ ♦

Case 2 n is odd

In this case, central inversion is not a symmetry of the figure, but (as we have seen) we may choose σ to be reflection in the xy-plane. Theorem 4.1 applies as before, but we can go further! We shall proceed with our analysis of the case $n = 3$.

In Example 4.1, we saw that $\Gamma^+(\text{DIH}_3)$ is generated by $r = (123)$ and $\rho = (NS)(23)$; thus the full symmetry group $\Gamma(\text{DIH}_3)$ has three generators:

$$r, \quad \rho \quad \text{and} \quad \sigma = (NS).$$

Various relations exists between them; in particular, we have already seen that σ commutes with all of $\Gamma^+(\text{DIH}_3)$, so that

$$\sigma r = r\sigma \quad \text{and} \quad \sigma\rho = \rho\sigma.$$

We shall now show that $\Gamma(\text{DIH}_3)$, in addition to being isomorphic to $D_3 \times C_2$, is also isomorphic to a *single* dihedral group!

In order to do this, let us examine the indirect symmetry σr.

Exercise 4.2

Determine the cyclic subgroup of $\Gamma(\text{DIH}_3)$ generated by σr. What is the order of σr?

We see that both σ and r can be expressed in terms of σr, so we can generate the whole group $\Gamma(\text{DIH}_3)$ using only the two generators σr and ρ. Moreover, we know the following relations involving these generators:

$$(\sigma r)^6 = \rho^2 = e.$$

This may remind you of the relations in a dihedral group! In fact all that is missing is a relation between σr and ρ.

Exercise 4.3

Using Table 4.2, or otherwise, calculate $\rho(\sigma r)\rho^{-1}$.

From the result of the last exercise, we see that $\rho(\sigma r)\rho^{-1} = \sigma r^{-1}$. This is almost enough to show that $\Gamma(\text{DIH}_3) \cong D_6$. If we can show that $\sigma r^{-1} = (\sigma r)^{-1}$, we shall be home and dry. Now

$$\begin{aligned}(\sigma r)^{-1} &= (r\sigma)^{-1} \quad \text{(since } \sigma \text{ and } r \text{ commute)} \\ &= \sigma^{-1} r^{-1} \\ &= \sigma r^{-1} \quad \text{(since } \sigma \text{ is of order 2)}.\end{aligned}$$

To sum up, we can write $\Gamma(\text{DIH}_3)$ in terms of generators and relations as

$$\Gamma(\text{DIH}_3) = \langle \sigma r, \rho : (\sigma r)^6 = e, \rho^2 = e, \rho(\sigma r)\rho^{-1} = (\sigma r)^{-1} \rangle,$$

which is isomorphic to the dihedral group D_6.

The general case is now easy.

Theorem 4.3

If n is odd, then $\Gamma(\text{DIH}_n)$ is isomorphic to the dihedral group D_{2n}.

Proof

We know that $\Gamma^+(\text{DIH}_n)$ is isomorphic to the dihedral group D_n, and is generated by a rotation r of order n and a rotation ρ of order 2. The rotation r has NS as axis, while the rotation ρ has as axis a horizontal line

Theorem 3.1.

joining the vertex 1 to the midpoint of the opposite side. Label the reflection in the xy-plane as σ; then σ commutes with both r and ρ. Moreover, the element σr generates a cyclic subgroup

$$\{e, \sigma r, r^2, \sigma r^3, \ldots\}.$$

There are $2n$ terms of the form $\sigma^i r^j$ ($i = 0, 1; j = 0, \ldots, n-1$); since they are all distinct, we see that this group has order $2n$, and contains the elements r and σ. Thus $\Gamma(\text{DIH}_n)$ is generated by σr and ρ, and $(\sigma r)^{2n} = \rho^2 = e$.

Moreover, we also have

$$\begin{aligned}
\rho(\sigma r)\rho^{-1} &= \sigma(\rho r \rho^{-1}) &&\text{(since σ and ρ commute)} \\
&= \sigma r^{-1} &&\text{(since $\Gamma^+(\text{DIH}_n) \cong D_n$)} \\
&= r^{-1}\sigma &&\text{(since σ and r commute)} \\
&= r^{-1}\sigma^{-1} &&\text{(since σ is of order 2)} \\
&= (\sigma r)^{-1}.
\end{aligned}$$

Thus $\Gamma(\text{DIH}_n)$ is generated by σr and ρ, subject to the relations

$$(\sigma r)^{2n} = e, \quad \rho^2 = e, \quad \rho(\sigma r)\rho^{-1} = (\sigma r)^{-1}.$$

This shows that $\Gamma(\text{DIH}_n)$ is indeed isomorphic to D_{2n}. ∎

4.4 The tetrahedron

We have seen that the rotation group $\Gamma^+(\text{TET})$ of the tetrahedron is isomorphic to A_4, and we know that the full symmetry group $\Gamma(\text{TET})$ can be written as $\Gamma^+(\text{TET}) \cup \sigma\Gamma^+(\text{TET})$, where σ is an indirect symmetry of the tetrahedron. Although central inversion is *not* a symmetry, this particular case is very easy because *any* reflective symmetry will suffice. Explicitly we choose reflection in the plane passing through 12 and the midpoint of 34 (see Figure 4.3).

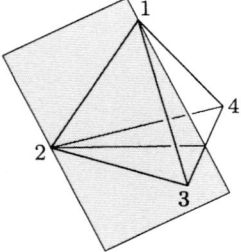

Figure 4.3

Thus σ can be written in cycle form as (34). As before, we could draw up tables of all the elements of $\Gamma^+(\text{TET})$ and $\Gamma^-(\text{TET})$, but in view of our prior knowledge this would be wasted effort.

Theorem 4.4

The full symmetry group $\Gamma(\text{TET})$ is isomorphic to S_4.

Algebraically this is clear, because a single 2-cycle together with A_4 generates the whole of S_4.

Geometrically it is easy to see that any permutation of the four vertices is achievable as long as we can use reflections of $\mathbb{R}^3$.

Note that S_4 is *not* isomorphic to $A_4 \times C_2$. Theorem 4.1 is not applicable here, as none of the indirect symmetries of the tetrahedron commute with all the direct symmetries.

4.5 The cube and octahedron

The full symmetry group $\Gamma(\text{CUBE})$ of the cube is immediately computable from earlier work. We know that $\Gamma^+(\text{CUBE})$ is isomorphic to S_4, but we also see that central inversion (which commutes with all symmetries) is a symmetry of the cube. Thus Theorem 4.1 tells us that

$$\Gamma(\text{CUBE}) \cong S_4 \times C_2.$$

As we know from the video programme that $\Gamma(\text{CUBE}) \cong \Gamma(\text{OCTA})$, the symmetry group of the octagon, we have the following theorem.

Theorem 4.5

The full symmetry group $\Gamma(\text{CUBE}) \cong \Gamma(\text{OCTA}) \cong S_4 \times C_2$.

4.6 The dodecahedron and icosahedron

Once again Theorem 4.1 comes to our assistance, because central inversion is a symmetry of the dodecahedron. Combining this with our knowledge from the video programme that

$$\Gamma^+(\text{DODECA}) \cong A_5 \quad \text{and} \quad \Gamma(\text{DODECA}) \cong \Gamma(\text{ICOSA}),$$

the latter being the symmetry group of the icosahedron, we obtain the following theorem.

Theorem 4.6

The full symmetry group $\Gamma(\text{DODECA}) \cong \Gamma(\text{ICOSA}) \cong A_5 \times C_2$.

In this section, we have begun our enumeration of indirect symmetry groups. In particular, Theorem 4.1 allows us to deduce that if central inversion belongs to G then G is isomorphic to $G^+ \times C_2$. The remaining results are summarized in Table 4.4.

By *indirect symmetry group* we mean a symmetry group that contains indirect symmetries.

Figure	Full symmetry group
dihedron DIH_n (n even)	$D_n \times C_2$
dihedron DIH_n (n odd)	$D_n \times C_2$ or D_{2n}
tetrahedron	S_4
cube	$S_4 \times C_2$
octahedron	$S_4 \times C_2$
dodecahedron	$A_5 \times C_2$
icosahedron	$A_5 \times C_2$

Table 4.4

This deals with most of the solid figures we have encountered so far, but we do not yet have a complete list of finite subgroups of $O(3)$. For this we proceed to Section 5.

5 A COMPLETE CATALOGUE

In this final section we justify our earlier work by showing that we have seen all possible finite *rotation* groups of $\mathbb{R}^3$, and then we identify the remaining finite symmetry groups.

5.1 Finite subgroups of $SO(3)$

We start with rotations, so let G be any finite subgroup of $SO(3)$. Since we are dealing with rotations, we know that the unit sphere is preserved (mapped to itself) by any element of G. Each rotation has an axis **a**, and such an axis meets the unit sphere in two diametrically opposite *poles* p and $-p$ (see Figure 5.1).

If you are short of time, you may wish to go straight to page 33.

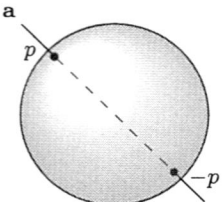

Figure 5.1

Consider a particular pole p. Any rotation ϕ of G does one of two things:

either the pole p is fixed, in which case ϕ is a rotation about the axis joining p to $-p$, as shown;

or the pole p is rotated to another point $\phi(p)$.

In the latter case, $\phi(p)$ is also a pole for some other rotation of G.

Before proving this, we look at an example.

Example 5.1

Recall that $\Gamma^+(\text{TET}) \cong A_4$. We can imagine the tetrahedron inscribed in the unit sphere, as shown in Figure 5.2.

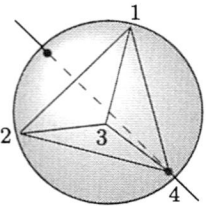

Figure 5.2

The rotation (123) has an axis passing through vertex 4 and the centre of the face 123. These two points determine a diameter of the sphere, and thus two poles. One of them is the vertex 4, the other is the antipodal point shown in the figure. Now any rotation ϕ of $\Gamma^+(\text{TET})$ does one of two things:

either the pole 4 is fixed, in which case ϕ is a rotation about the axis shown;

or the pole 4 is rotated to one of the vertices 1, 2 or 3.

The content of the theorem is that in the latter case G must also contain a rotation about the image point. We know this for $\Gamma^+(\text{TET})$ from earlier work, but the theorem applies to *any* finite subgroup of $SO(3)$. ◆

> **Theorem 5.1**
>
> Let G be any finite subgroup of $SO(3)$ and let ϕ be any rotation in G. If ϕ does not fix the pole p, then $\phi(p)$ is also a pole for some other rotation of G.

Proof

Let r_p be any rotation in G with axis $\{p, -p\}$, and consider the effect of $\phi r_p \phi^{-1}$ on $\phi(p)$. Certainly $\phi r_p \phi^{-1}$ is in G, and

$$\phi r_p \phi^{-1} \phi(p) = \phi r_p(p) = \phi(p).$$

Thus $\phi r_p \phi^{-1}$ is a rotation in G which fixes $\phi(p)$ and so has $\phi(p)$ as a pole. ∎

Throughout this section, 'the axis $\{p, -p\}$' means 'the axis through the points p and $-p$'.

In summary, a typical rotation either fixes a pole or moves it to another pole.

This means that the group G acts on the set P of all poles of G. In the remainder of this section we shall use orbit–stabilizer theory to deduce that only certain groups G are possible. We present the argument in an informal fashion without appealing explicitly to standard theorems.

The stabilizer of the pole p is precisely the cyclic subgroup of G consisting of rotations about the axis $\{p, -p\}$. Denote the order of this stabilizer by n_p; we shall refer to n_p as the *order* of the pole p.

If another pole q is in the orbit of p, then its stabilizer is isomorphic to that of p, and so

$$n_q = n_p.$$

Example 5.2

Consider $\Gamma^+(\text{CUBE})$. Let p be the pole determined by the face 1234, as shown in Figure 5.3.

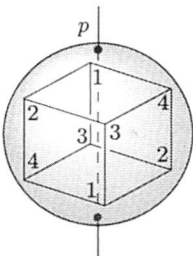

Figure 5.3

Then $\text{Stab}(p)$ is the cyclic subgroup of rotations about the centre of face 1234.

The orbit $\text{Orb}(p)$ is the set of six poles determined by the six faces of the cube. For any such centre, the stabilizer is the cyclic subgroup of rotations about that centre. All the stabilizers have order 4, so here $n_p = 4$. ◆

The poles are on the unit sphere, *not* on the cube.

Example 5.3

Again in $\Gamma^+(\text{CUBE})$, consider the pole p defined by one of the vertices — say the vertex labelled 4 in Figure 5.4.

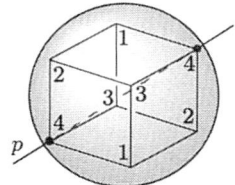

Figure 5.4

Now Orb(p) is precisely the set of eight vertices. Stab(p) is the cyclic group of rotations about the diagonal through p. These stabilizers are all isomorphic to the cyclic group of order 3, so $n_p = 3$ in this case. ♦

Note that Stab(p) contains $n_p - 1$ non-trivial rotations.

It is an immediate consequence of the Orbit–stabilizer Theorem that the number of poles in the orbit of p is $|G|/n_p$.

By the Orbit–stabilizer Theorem, $|\text{Orb}(p)| = |G|/|\text{Stab}(p)|$.

Example 5.4

For the group Γ^+(DODECA) there are three types of pole:

(a) those determined by the centre of a face;
(b) those determined by the midpoint of an edge;
(c) the vertices.

For type (a) we have as stabilizer the cyclic group of order 5, and the number of poles in the orbit is $60/5 = 12$ (i.e. those determined by all twelve face centres).

Since Γ^+(DODECA) $\cong A_5$, it follows that the order of this group is 60.

For type (b) we have as stabilizer the cyclic group of order 2, and the number of poles in the orbit is $60/2 = 30$ (i.e. those determined by every edge midpoint).

For type (c) we have as stabilizer the cyclic group of order 3, and the number of poles in the orbit is $60/3 = 20$ (i.e. every vertex). ♦

Now every non-trivial rotation belongs to the stabilizer of some pole.

Counting up over *all* the orbits, we have a total of

$$\sum_{\text{orbits}} (n_p - 1) \frac{|G|}{n_p}$$

non-trivial rotations. This counts every non-trivial rotation *twice*, because each rotation is associated with *two* poles.

Example 5.5

For the group Γ^+(DODECA) there are just three orbits, so the above sum contains three terms.

For type (a) we have $n_p = 5$, giving $4 \times 60/5 = 48$.

Type (b) has $n_p = 2$, so the term here is $1 \times 60/2 = 30$.

Finally type (c) has $n_p = 3$, giving $2 \times 60/3 = 40$.

Thus the total sum is 118, which is indeed $2 \times (60 - 1)$. ♦

So we have

$$\sum_{\text{orbits}} (n_p - 1) \frac{|G|}{n_p} = 2(|G| - 1),$$

which simplifies to

$$\sum_{\text{orbits}} \left(1 - \frac{1}{n_p}\right) |G| = 2(|G| - 1)$$

or (dividing through by $|G|$)

$$\sum_{\text{orbits}} \left(1 - \frac{1}{n_p}\right) = 2\left(1 - \frac{1}{|G|}\right). \tag{5.1}$$

This equation provides the result we are after, for $|G|$ and n_p are both at least 2. Now the fact that $|G| \geq 2$ means that

$$0 < 1/|G| \leq \frac{1}{2},$$

and so

$$\frac{1}{2} \leq 1 - \frac{1}{|G|} < 1.$$

Thus the right-hand side of Equation 5.1 has a value lying in the range $[1, 2[$.

However, since $n_p \geq 2$, we have

$$\frac{1}{n_p} \leq \frac{1}{2},$$

and so

$$\frac{1}{2} \leq 1 - \frac{1}{n_p} < 1.$$

Thus each term on the left-hand side of Equation 5.1 lies in the range $[\frac{1}{2}, 1[$.

If the left-hand side contained only one term, it could not reach the minimum value 1 of the right-hand side. Therefore the left-hand side must have at least two terms. However, if it had four or more terms, then its minimum value would be 2, which is greater than the range of maximum permitted values on the right.

Thus the sum on the left can produce only two or three terms, and so we have the following theorem.

> **Theorem 5.2**
>
> If G is a non-trivial finite subgroup of $SO(3)$, with set of poles P, then the action of G on P can produce only two or three orbits.

All that remains is to analyse these possibilities separately.

> We can take it that $|G| \geq 2$ because, if G were trivial, there would be no poles to count!

Case 1 Two orbits

In this case Equation 5.1 becomes

$$\left(1 - \frac{1}{n_p}\right) + \left(1 - \frac{1}{n_q}\right) = 2 - \frac{2}{|G|},$$

where n_p and n_q are the orders of the poles in each orbit. Thus

$$\frac{1}{n_p} + \frac{1}{n_q} = \frac{2}{|G|}. \tag{5.2}$$

Now each n is the order of a subgroup of G, and so must divide $|G|$. Thus, for each n, we have $n \leq |G|$, and so

$$\frac{1}{n} \geq \frac{1}{|G|}.$$

This in turn means that

$$\frac{1}{n_p} + \frac{1}{n_q} \geq \frac{2}{|G|}.$$

Therefore Equation 5.2 has a solution only if $n_p = n_q = |G|$.

The size of each orbit is one, and there are therefore just two (diametrically opposite) poles. The group G is a cyclic group of rotations about the axis through the two poles.

Case 2 Three orbits

In this case Equation 5.1 becomes

$$\left(1-\frac{1}{n_p}\right)+\left(1-\frac{1}{n_q}\right)+\left(1-\frac{1}{n_r}\right)=2-\frac{2}{|G|},$$

or

$$1+\frac{2}{|G|}=\frac{1}{n_p}+\frac{1}{n_q}+\frac{1}{n_r}. \tag{5.3}$$

Exercise 5.1

Show that if all the ns are greater than 2 then Equation 5.3 has no solution.

Suppose that $n_p = 2$. We now have

$$\frac{1}{2}+\frac{2}{|G|}=\frac{1}{n_q}+\frac{1}{n_r}.$$

One possibility is that n_q is also 2, in which case we have $n_r = \frac{1}{2}|G| = k$, say. Our first solution is thus

$$n_p = 2, \quad n_q = 2, \quad n_r = k.$$

If $k \geq 2$, this case corresponds to the rotation group of a dihedron with k equatorial vertices. We can see this informally since there are just two poles of order k. These are the opposite pair N, S and the stabilizer is the cyclic group of rotations about NS. Any other pole is of order 2 and has k poles in its orbit. These can be equally spaced around the equator and we see that we have constructed a dihedron with k equatorial vertices.

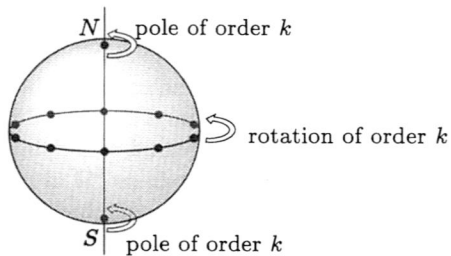

Figure 5.5

Next suppose that $n_q \neq 2$.

Exercise 5.2

Show that if both n_q and n_r are greater than 3, then Equation 5.3 has no solution.

So we may assume that $n_q = 3$. Equation 5.3 now becomes

$$\frac{1}{6}+\frac{2}{|G|}=\frac{1}{n_r}.$$

This clearly means that n_r is less than 6, and so must be 3, 4 or 5. We consider these possibilities in turn.

Firstly, if

$$n_p = 2, \quad n_q = 3, \quad n_r = 3,$$

then $|G| = 12$ and we have $\Gamma^+(\text{TET})$. You can visualize this because a pole of order 3 has four poles in its orbit. Any rotation which fixes it permutes the other three and so these four poles must form the vertices of a regular tetrahedron. The remaining four poles of order 3 are those determined by the centres of faces, and the six poles of order 2 are those determined by the midpoints of edges.

Secondly, if
$$n_p = 2, \quad n_q = 3, \quad n_r = 4,$$
then $|G| = 24$ and we have $\Gamma^+(\text{CUBE})$. To prove this is slightly more tricky, so we content ourselves with noticing that the poles of order 2 (twelve in the orbit) correspond to the twelve edges, those of order 3 (eight in the orbit) to the eight vertices, and the rotations of order 4 (six in the orbit) to the six faces.

Finally, if
$$n_p = 2, \quad n_q = 3, \quad n_r = 5,$$
then $|G| = 60$ and we have $\Gamma^+(\text{DODECA})$. Again this requires proof, but we notice that there are 20 poles of order 3. It turns out that these form the vertices of a regular dodecahedron. The 30 poles of order 2 correspond to the opposite pairs of edge midpoints, and the 12 poles of order 5 correspond to the centres of the faces.

We can summarize the results of this subsection in the following theorem.

If you decided to skip the first part of Section 5, this is where you should climb aboard again!

Theorem 5.3

A finite subgroup of $SO(3)$ is isomorphic to one of the following groups:

(a) the cyclic group C_n;

(b). the dihedral group D_n;

(c) the alternating group A_4;

(d) the symmetric group S_4;

(e) the alternating group A_5.

Although we have not proved rigorously that there are only five regular solids, we have certainly shown that no other possible regular solid can have a direct symmetry group other than the ones we have listed. Indeed, if S is a regular solid, then its direct symmetry group must have poles in three orbits, corresponding to face centres, edge midpoints and vertices. The above analysis goes some way towards justifying that we have met them all.

5.2 The remaining finite subgroups of $O(3)$

Theorem 4.1 takes care of many of the finite subgroups of $O(3)$ containing indirect isometries. With the aid of the following theorem we shall account for the rest.

Theorem 5.4

Let G be a subgroup of $O(3)$ containing indirect isometries, so that $G = H \cup \sigma H$, where H is a subgroup of $SO(3)$ and σ is an indirect isometry in G. If σ_O is not an element of G, then

$$G^* = H \cup \sigma_O \sigma H$$

is a subgroup of $SO(3)$ and G is isomorphic to G^*.

Proof

Since the product of two indirect isometries is direct, we see that $\sigma_O \sigma$ is direct, and so $G^* \subseteq SO(3)$.

We now need to check the *closure, inverses* and *identity* properties. It is helpful to recall that σ_O commutes with every element of $O(3)$.

Closure

To see that G^* is closed, we need check only the effects of $\sigma_O \sigma H$. Now let $h_1, h_2 \in H$. Then
$$h_1(\sigma_O \sigma h_2) = \sigma_O(h_1 \sigma h_2)$$
$$= \sigma_O(\sigma h_3) \in G^* \quad \text{(since } h_1 \sigma h_2 \text{ is an indirect isometry}$$
$$\sigma h_3 \text{ of } G \text{ for some } h_3 \in H).$$

Also,
$$(\sigma_O \sigma h_2) h_1 = \sigma_O(\sigma h_2 h_1) \in G^*,$$
and
$$(\sigma_O \sigma h_1)(\sigma_O \sigma h_2) = (\sigma_O^2 \sigma h_1)(\sigma h_2)$$
$$= \sigma h_1 \sigma h_2 \in H \subseteq G^* \quad \text{(since it is direct)}.$$

Therefore, G^* is closed.

Inverses

The inverse of $\sigma_O(\sigma h_1)$ is
$$(\sigma h_1)^{-1} \sigma_O^{-1} = (\sigma h_1)^{-1} \sigma_O$$
$$= \sigma_O(\sigma h_1)^{-1},$$
which again lies in G^*.

Identity

Clearly, $e \in H \subseteq G^*$.

So G^* is certainly a subgroup of $SO(3)$.

Finally, we show that G is isomorphic to G^*.

We define an isomorphism ϕ from G to G^* by:
$$\phi(h_1) = h_1, \quad \text{if } h_1 \in H;$$
$$\phi(\sigma h_2) = \sigma_O \sigma h_2, \quad \text{if } h_2 \in H.$$

The mapping ϕ is a bijection, and obviously satisfies the homomorphism property as far as elements of H are concerned. Thus it remains to check this property for elements of σH composed with elements of H, and elements of σH composed with elements of σH.

Now
$$\phi((\sigma h_2) h_1) = \phi(\sigma(h_2 h_1)) = \sigma_O \sigma(h_2 h_1) = (\sigma_O \sigma h_2) h_1 = \phi(\sigma h_2) \phi(h_1),$$
and
$$\phi(h_1(\sigma h_2)) = \phi(h_1 \sigma h_2) = \sigma_O(h_1 \sigma h_2) \quad \text{(since } h_1 \sigma h_2 \text{ must be indirect)}$$
$$= h_1(\sigma_O \sigma h_2)$$
$$= \phi(h_1) \phi(\sigma h_2).$$

Finally,
$$\phi((\sigma h_1)(\sigma h_2)) = \phi(\sigma h_1 \sigma h_2)$$
$$= \sigma h_1 \sigma h_2 \quad \text{(since } \sigma h_1 \sigma h_2 \text{ must be direct)}$$
$$= \sigma_O^2(\sigma h_1 \sigma h_2) \quad \text{(since } \sigma_O \text{ has order 2)}$$
$$= (\sigma_O \sigma h_1)(\sigma_O \sigma h_2)$$
$$= \phi(\sigma h_1) \phi(\sigma h_2).$$

It follows that $G \cong G^*$. ∎

Note that, as ϕ maps some indirect isometries to direct isometries, it is *not* true that G is *conjugate* to G^*.

Now we know from Theorem 2.5 that any finite subgroup of $O(3)$ containing indirect isometries has a decomposition as

$$G = G^+ \cup \sigma G^+,$$

where σ is any of the indirect isometries of G and G^+ is one of the types listed in Theorem 5.3. Moreover,

- if there is a σ commuting with all elements of G^+, then $G \cong G^+ \times C_2$ (by Theorem 4.1);
- if $\sigma_O \notin G$, then G is itself isomorphic to one of the types in Theorem 5.3, and contains G^+ as a subgroup of index 2 (by Theorem 5.4).

These two possibilities cover all cases where G contains indirect isometries, since σ_O commutes with all elements of $O(3)$, let alone all elements of G^+. However, they are not mutually exclusive — we have seen that if $G^+ = \Gamma^+(\text{DIH}_n)$ and n is odd, then $\sigma_O \notin G$ but there is nevertheless a σ commuting with all elements of G^+ (namely reflection in the xy-plane). Thus G has two expressions: it is isomorphic *both* to $D_n \times C_2$ *and* to D_{2n}.

In the final subsection, we catalogue all the possibilities that arise in this way out of Theorems 2.5, 4.1, 5.3 and 5.4, giving in each case a geometric figure whose symmetry group is isomorphic to the group in question. This provides a visual summary of the section, and indeed the whole unit.

5.3 A complete catalogue

In order to give examples of all possible finite subgroups of $O(3)$ containing indirect isometries, we need certain basic figures to work with. There are many ways in which such figures can be constructed, but perhaps the simplest is based on the *prism*.

By a prism we mean a thickened copy of a regular polygon. Specifically, if the regular n-gon is given depth we obtain an n-prism. Thus, Figure 5.6 shows a 6-prism.

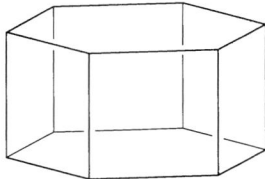

Figure 5.6

Now we start the cataloguing process. At the end of this section you should be able to:

(a) give an example of a geometric figure which has a given symmetry group;

(b) write down the symmetry group of any of the figures dealt with below.

Let us summarize our knowledge. To begin with, we know from Theorem 2.5 that if G contains any indirect symmetries then it has a decomposition into two cosets:

$$G = G^+ \cup \sigma G^+.$$

Next, we know from Theorem 5.3 that G^+ is isomorphic to one of just five types (see Figure 5.7).

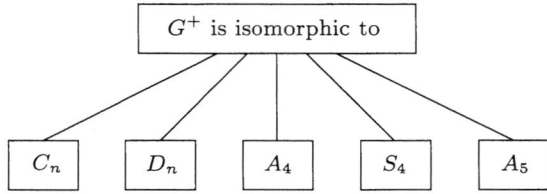

Figure 5.7

Finally, Theorems 4.1 and 5.4 allow us to express $G^+ \cup \sigma G^+$ as being isomorphic *either* to $G^+ \times C_2$ *or* to G^* where G^* is also isomorphic to one of the five types; in some cases *both* of these alternatives are available.

We shall examine the possibilities systematically, going through each possibility for G^+, then considering (for each of these possibilities) the case where $\sigma_O \in G$ and $\sigma_O \notin G$.

We start by considering the possibility that G^+ is the cyclic group C_n. We must consider whether or not σ_O belongs to G.

If $\sigma_O \in G$, then $G \cong C_n \times C_2$, and we now need to know the parity of n.   Compare with Subsection 4.1.

For even values of n, this group is the symmetry group of a *marked prism*; that is, a prism which is marked in a particular way. Figure 5.8 shows a marked 6-prism.

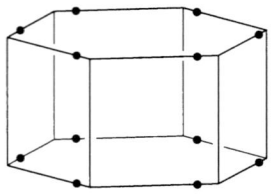

Figure 5.8

The marking points are placed near each vertex, and the effect is to prevent any rotation from interchanging the top and bottom faces. (Such a rotation would have a horizontal axis, and you can see that it would not preserve the marking.) Thus the rotation group consists of just those rotations about the vertical axis, and these form a cyclic group, in this case C_6. So the full symmetry group of this figure is $C_6 \times C_2$, and a similar argument works for any even value of n.

For odd values of n, we have a slight problem, because central inversion is not a symmetry of such a marked n-prism. To cope with this, we use a $2n$-prism and place $2n$ marking points alternately on top and bottom edges (see Figure 5.9).

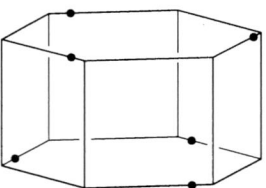

Figure 5.9

This still has the effect of preventing rotational symmetry about a horizontal axis, and the rotation group is just the cyclic group C_n (in this particular case C_3). Central inversion ensures that the full symmetry group is $C_n \times C_2$ in general, and $C_3 \times C_2$ for this particular group.

The next possibility we must deal with is that central inversion is not an element of G. In this case we appeal to Theorem 5.4 and deduce that G is itself isomorphic to one of the rotation groups. So we examine the list of finite subgroups of $SO(3)$, looking for groups which have C_n as a subgroup of index 2. In fact there are only two possibilities: C_{2n} and D_n.

First, suppose that $G \cong C_{2n}$: once again our use of prisms is governed by the parity of n.

This time, for odd values of n we can use the marked n-prism, such as that shown in Figure 5.10, where $n = 5$.

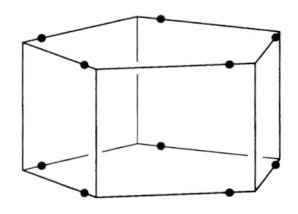

Figure 5.10

The symmetry group of this marked 5-prism is isomorphic to C_{10}. A generator g is represented by reflection in the xy-plane composed with rotation through $2\pi/5$ about the vertical axis. Clearly g^2 is rotation through $4\pi/5$ about the vertical axis, and so on. The rotations form a subgroup C_5 of index 2. A similar argument works for any odd n.

Note that if we take σ to be reflection in the horizontal plane bisecting the prism, then σ commutes with G^+, so we *also* obtain $G \cong C_n \times C_2$. But this is not surprising: we know from Theorem 5.1 of *Unit GR2* that if m and n are coprime then $C_m \times C_n \cong C_{mn}$.

Now if n is even, central inversion is a symmetry of a marked n-prism, but is not in G, and so we use a $2n$-prism, with $2n$ marking points placed alternately on upper and lower faces.

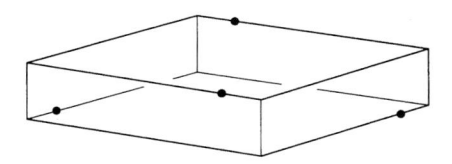

Figure 5.11

Figure 5.11 shows the case $n = 2$. The resulting 4-prism has symmetry group isomorphic to C_4. This is generated by the symmetry g, obtained by composing reflection in the xy-plane with a rotation through $\pi/2$ about the vertical axis. Clearly g^2 is rotation about the vertical axis through π, showing C_2 as a rotation subgroup of index 2. The same sort of argument works for any even value of n.

To complete this line of enquiry, we need to deal with the case $G^+ \cong C_n$, $G \cong D_n$.

But this is easy: we take an n-pyramid.

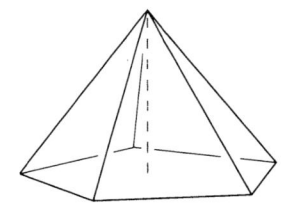

Figure 5.12

Figure 5.12 shows a 5-pyramid, which is self-explanatory. Its symmetry group is isomorphic to D_5. Rotations about the vertical axis form a subgroup isomorphic to C_5, and reflection in a vertical plane passing through the apex and through one of the five vertices in the xy-plane is the remaining generator.

Figure 5.13 summarizes the possibilities that arise if G^+ is isomorphic to C_n.

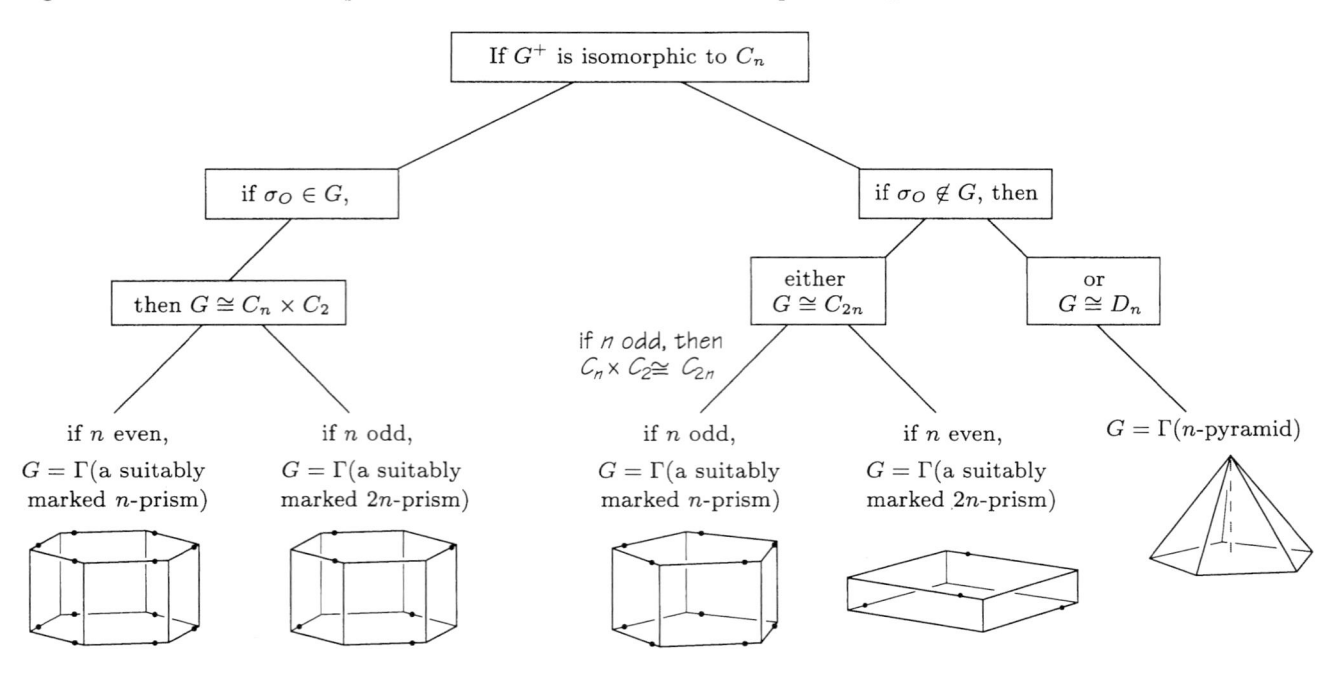

```
                    If G⁺ is isomorphic to Cₙ
                   /                          \
           if σ_O ∈ G,                    if σ_O ∉ G, then
              |                            /            \
       then G ≅ Cₙ × C₂                either           or
         /        \                   G ≅ C₂ₙ          G ≅ Dₙ
                             if n odd, then
                             Cₙ × C₂ ≅ C₂ₙ
```

| if n even, | if n odd, | if n odd, | if n even, | $G = \Gamma(n\text{-pyramid})$ |
| $G = \Gamma$(a suitably marked n-prism) | $G = \Gamma$(a suitably marked $2n$-prism) | $G = \Gamma$(a suitably marked n-prism) | $G = \Gamma$(a suitably marked $2n$-prism) | |

Figure 5.13

We have now dealt with all the cases arising from the possibility that $G^+ = C_n$. The second branch of the overall tree of possibilities arises from the assumption that $G^+ = D_n$. Once again we shall need to take account of the presence or absence of σ_O.

So let us first assume that σ_O is in G, which is therefore isomorphic to $D_n \times C_2$.

For even values of n we have seen earlier that the dihedron DIH_n will do. In keeping with our general use of prisms, we could equally well use the n-prism.

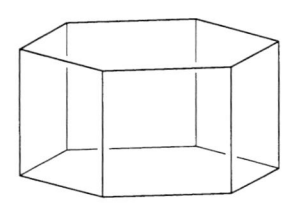

Figure 5.14

Figure 5.14 shows a 6-prism, with a symmetry group isomorphic to $D_6 \times C_2$. The symmetries which preserve the upper face form a subgroup isomorphic to D_6, and combination with central inversion provides the rest.

If n is odd, we cannot use an n-prism because it does not have σ_O as a symmetry. So we take a $2n$-prism and mark top and bottom edges alternately (see Figure 5.15, where $n = 3$).

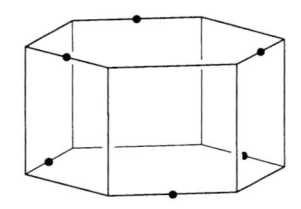

Figure 5.15

Notice, however, that the markings are now at the *midpoints* of the edges. The reason is that we wish to have a rotation group isomorphic to D_n, rather than C_n as earlier. Putting the markings at edge midpoints ensures this by allowing rotations about horizontal axes as symmetries. (A typical one is a rotation through π about a line joining the midpoints of two opposite vertical edges.)

To complete this second branch, we now assume that σ_O is not in G. Checking through the list in Theorem 5.3 we see that G must be isomorphic to D_{2n}.

For odd values of n we have already seen that the dihedron DIH_n will do — or, in our present context, the n-prism.

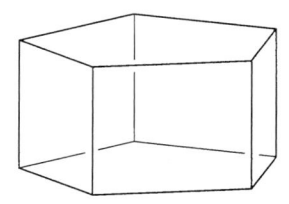

Figure 5.16

Figure 5.16 shows a 5-prism, with symmetry group isomorphic to D_{10}. There are two generators, namely:

- a composite of reflection in the xy-plane with rotation through $2\pi/5$ about the vertical axis; this generates a cyclic subgroup of order 10;
- rotation through π about a horizontal axis joining a midpoint of one of the vertical edges to the centre of the opposite face.

As before, if we take σ to be reflection in the horizontal bisecting plane, then σ commutes with G^+, so we *also* obtain $G \cong D_n \times C_2$. Thus, for odd n, the symmetry groups of the n-prism and the marked $2n$-prism are isomorphic (but not conjugate).

Finally, if n is even, then we mark a $2n$-prism at alternate midpoints.

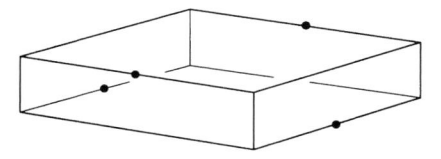

Figure 5.17

Figure 5.17 shows $n = 2$. Once again there are two generators of the symmetry group:

- a composite of reflection in the xy-plane with rotation through $\pi/2$ about the vertical axis: this generates a cyclic subgroup of order 4;
- rotation through π about a horizontal axis joining midpoints of opposite vertical edges.

Figure 5.18 summarizes the possibilities that arise if G^+ is isomorphic to C_n.

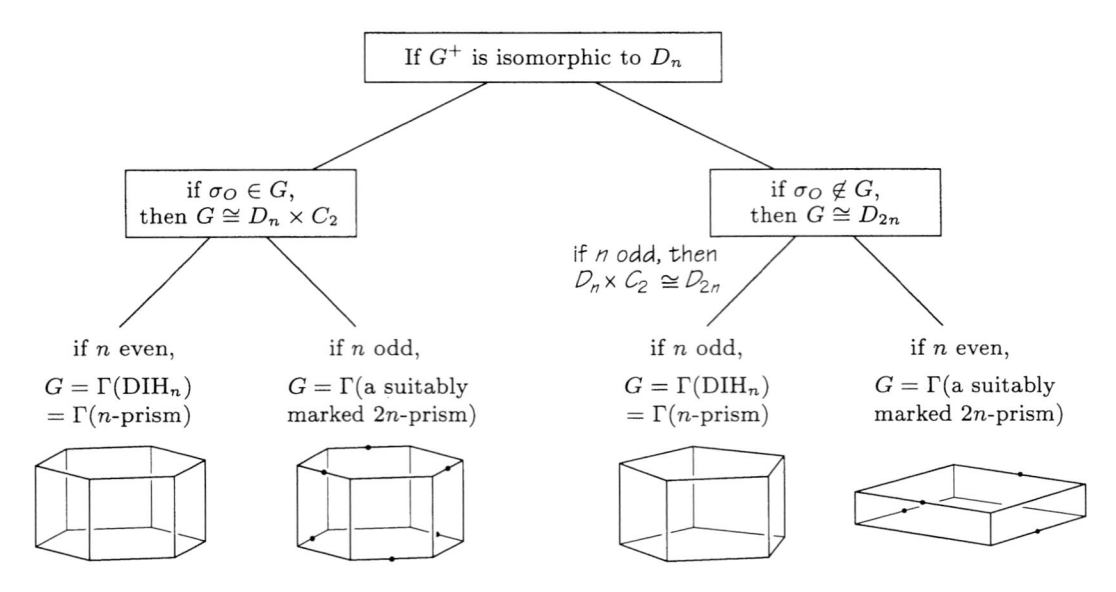

```
                  If G+ is isomorphic to Dn
                   /                    \
     if σ_O ∈ G,                          if σ_O ∉ G,
     then G ≅ Dn × C2                     then G ≅ D2n
       /          \          if n odd, then     /         \
                              Dn × C2 ≅ D2n
   if n even,   if n odd,            if n odd,      if n even,
   G = Γ(DIHn)  G = Γ(a suitably     G = Γ(DIHn)    G = Γ(a suitably
   = Γ(n-prism) marked 2n-prism)     = Γ(n-prism)   marked 2n-prism)
```

Figure 5.18

All that remains is to check the final three branches, shown in Figure 5.7, starting with A_4.

To find a solid with symmetry group isomorphic to $A_4 \times C_2$, we use the fact that A_4 is a subgroup of S_4, and mark a cube.

Figure 5.19 has a total of three rotations of order 2 about axes through the centres of opposite faces, and eight further rotations of order 3 about axes through opposite corners. It has no rotations of order 4, and hence the direct symmetry group must be isomorphic to A_4. Next, notice that central inversion is a symmetry of this figure, so the full symmetry group is isomorphic to $A_4 \times C_2$, as required.

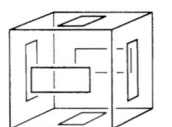

Figure 5.19

We have seen that the tetrahedron realizes the case when central inversion is not a symmetry. The full symmetry group is isomorphic to S_4, because no other subgroup of $SO(3)$ has A_4 as a subgroup of index 2.

Figure 5.20 summarizes the possibilities that arise if G^+ is isomorphic to A_4.

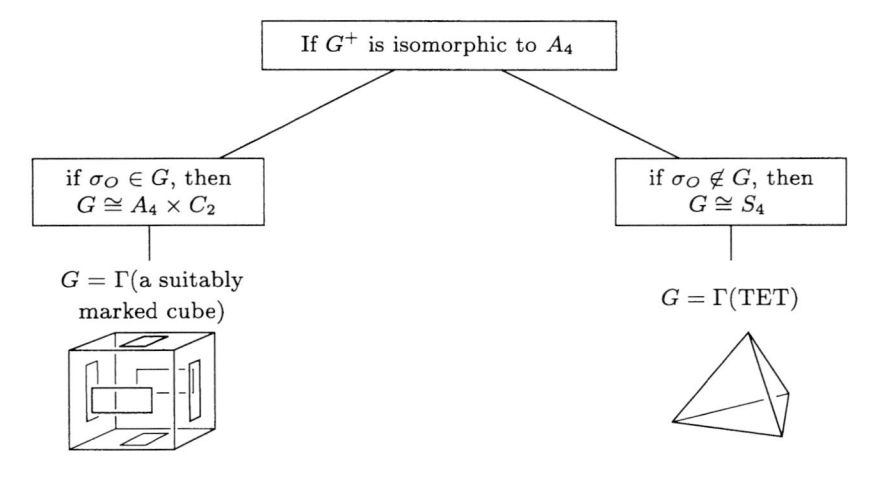

```
                If G+ is isomorphic to A4
                 /                     \
     if σ_O ∈ G, then          if σ_O ∉ G, then
     G ≅ A4 × C2                G ≅ S4
         |                          |
   G = Γ(a suitably             G = Γ(TET)
   marked cube)
```

Figure 5.20

Now for the penultimate branch, namely the case where G^+ is isomorphic to S_4. There is no direct symmetry group in the list of Theorem 5.3 which contains S_4 as a subgroup of index 2, and so we are forced to conclude that in this case we must have

$$G \cong \Gamma(\text{CUBE}) \cong S_4 \times C_2.$$

```
┌─────────────────────────────────┐
│ If $G^+$ is isomorphic to $S_4$ │
└─────────────────────────────────┘
                │
        ┌───────────────────┐
        │ $G \cong S_4 \times C_2$ │
        └───────────────────┘
                │
        $G = \Gamma(\text{CUBE})$
```

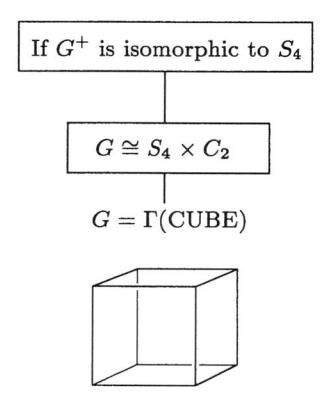

Figure 5.21

And now the final step, corresponding to the case where G^+ is isomorphic to A_5. Once again the list following Theorem 5.3 does not contain a group which has A_5 as a subgroup of index 2. Thus we must have

$$G \cong \Gamma(\text{DODECA}) \cong A_5 \times C_2.$$

```
┌─────────────────────────────────┐
│ If $G^+$ is isomorphic to $A_5$ │
└─────────────────────────────────┘
                │
        ┌───────────────────┐
        │ $G \cong A_5 \times C_2$ │
        └───────────────────┘
                │
        $G = \Gamma(\text{DODECA})$
```

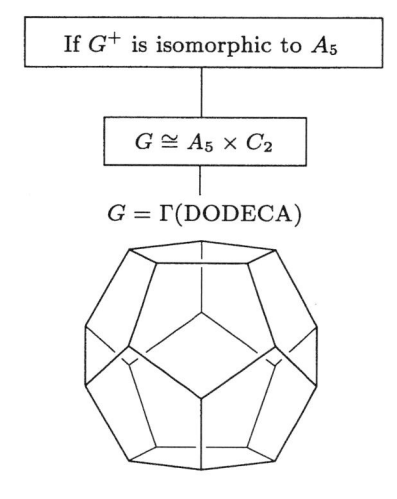

Figure 5.22

This completes our search, and the unit!

SOLUTIONS TO THE EXERCISES

Solution 1.1

(a) Since G is a group, $r[\theta_{j+1}](r[\theta_j])^{-1} \in G$ $(j = 1, 2, \ldots, n-1)$. But this element is $r[\theta_{j+1} - \theta_j]$. Thus, $\theta_{j+1} - \theta_j$ must be one of the θ_i. But, if $\theta_1 < \theta_{j+1} - \theta_j$, then $\theta_j + \theta_1$ lies between θ_j and θ_{j+1}. Since $r[\theta_j + \theta_1] \in G$, this contradicts the way the elements of G were listed. Therefore, $\theta_{j+1} - \theta_j = \theta_1$ $(j = 1, 2, \ldots, n-1)$.

(b) From part (a), it follows that
$$\theta_j = j\theta_1 \quad (j = 2, 3, \ldots, n).$$
This equation automatically holds for $j = 1$. Thus,
$$r[\theta_j] = (r[\theta_1])^j \quad (j = 1, 2, \ldots, n),$$
and so G is generated by $r[\theta_1]$. Since $\theta_n = 2\pi$, it also follows that $\theta_1 = 2\pi/n$.

Solution 1.2

We have
$$sr = r^{n-1}s.$$
Multiplying both sides on the right by s^{-1}, we obtain
$$\begin{aligned}srs^{-1} &= r^{n-1}(ss^{-1})\\ &= r^{n-1} \quad \text{(since } ss^{-1} = e\text{)}\\ &= r^n r^{-1}\\ &= r^{-1} \quad \text{(since } r^n = e\text{)},\end{aligned}$$

as required.

Note that s has order 2, so $s^{-1} = s$, but we write s^{-1} in order to express r^{-1} as a conjugate of r.

Solution 1.3

The identity does not belong to G^-.

Alternatively, the product of two elements of G^- is in G^+, so G^- is not closed.

Solution 1.4

$\widetilde{G}$ contains
$$\begin{aligned}r[-\theta]\, q[\theta]\, r[\theta] &= q[\theta + \tfrac{1}{2}(-\theta)]\, r[\theta] \quad \text{(by Equation 4 of the Isometry Toolkit)}\\ &= q[\tfrac{1}{2}\theta]\, r[\theta]\\ &= q[0] \quad \text{(by Equation 5 of the Isometry Toolkit)}.\end{aligned}$$

Solution 1.5

The matrix representing central inversion in $\mathbb{R}^2$ is $\begin{bmatrix} -1 & 0 \\ 0 & -1 \end{bmatrix}$, which is the matrix of a rotation through π.

Solution 2.1

We argue just as in the two-dimensional case. The distance between $\phi(\mathbf{a})$ and $\phi(\mathbf{b})$ is the same as that between $\mathbf{a}$ and $\mathbf{b}$. Thus
$$(\phi(\mathbf{a}) - \phi(\mathbf{b})) \cdot (\phi(\mathbf{a}) - \phi(\mathbf{b})) = (\mathbf{a} - \mathbf{b}) \cdot (\mathbf{a} - \mathbf{b}),$$
giving
$$\phi(\mathbf{a}) \cdot \phi(\mathbf{a}) + \phi(\mathbf{b}) \cdot \phi(\mathbf{b}) - 2\phi(\mathbf{a}) \cdot \phi(\mathbf{b}) = \mathbf{a} \cdot \mathbf{a} + \mathbf{b} \cdot \mathbf{b} - 2\mathbf{a} \cdot \mathbf{b}.$$
Now
$$\phi(\mathbf{a}) \cdot \phi(\mathbf{a}) = \mathbf{a} \cdot \mathbf{a} \quad \text{and} \quad \phi(\mathbf{b}) \cdot \phi(\mathbf{b}) = \mathbf{b} \cdot \mathbf{b},$$
so
$$\phi(\mathbf{a}) \cdot \phi(\mathbf{b}) = \mathbf{a} \cdot \mathbf{b},$$
as required.

Solution 2.2

Certainly the transformation

$$\begin{bmatrix} \cos 2\theta & \sin 2\theta & 0 \\ \sin 2\theta & -\cos 2\theta & 0 \\ 0 & 0 & 1 \end{bmatrix}$$

fixes the z-axis. Its effect on the xy-plane is that of the transformation

$$\begin{bmatrix} \cos 2\theta & \sin 2\theta \\ \sin 2\theta & -\cos 2\theta \end{bmatrix},$$

which is reflection in the line $x \sin\theta - y \cos\theta = 0$. Hence the result.

Solution 2.3

Any reflection of $\mathbb{R}^3$ has a plane of fixed points, but central inversion has no fixed points (except the origin).

Solution 2.4

We may choose our coordinates in such a way that the plane is the xy-plane and the perpendicular axis is the z-axis. Then reflection in the plane is represented by the matrix

$$\begin{bmatrix} 1 & 0 & 0 \\ 0 & 1 & 0 \\ 0 & 0 & -1 \end{bmatrix}$$

and rotation through π about the axis is given by

$$\begin{bmatrix} -1 & 0 & 0 \\ 0 & -1 & 0 \\ 0 & 0 & 1 \end{bmatrix}.$$

Composing the two gives

$$\begin{bmatrix} -1 & 0 & 0 \\ 0 & -1 & 0 \\ 0 & 0 & 1 \end{bmatrix} \begin{bmatrix} 1 & 0 & 0 \\ 0 & 1 & 0 \\ 0 & 0 & -1 \end{bmatrix} = \begin{bmatrix} -1 & 0 & 0 \\ 0 & -1 & 0 \\ 0 & 0 & -1 \end{bmatrix},$$

which represents central inversion.

Alternatively, you could trace the effect of the composite on $\mathbf{i}, \mathbf{j}$, and $\mathbf{k}$.

Solution 2.5

The matrix representing σ_O is

$$\begin{bmatrix} -1 & 0 & 0 \\ 0 & -1 & 0 \\ 0 & 0 & -1 \end{bmatrix} = -\mathbf{I}.$$

If $\mathbf{A}$ is the matrix representing any other element of $O(3)$, then

$$(-\mathbf{I})\mathbf{A} = \mathbf{A}(-\mathbf{I}) = -\mathbf{A}.$$

Thus σ_O commutes with every element of $O(3)$.

Solution 2.6

If G contains an indirect symmetry σ and a non-trivial rotation r, then it also contains another indirect symmetry σr. It follows that if G contains only one indirect symmetry, then $G \cong D_1$.

Of course, $D_1 \cong C_2$, so we could equally well say that $G \cong C_2$.

Solution 3.1

Keep N and S pointwise fixed.

Solution 4.1

The elements of $\sigma_O \Gamma^+(\text{DIH}_4)$ are as follows.

Element	Name
$(NS)(13)(24)$	σ_O
$(NS)(1432)$	$\sigma_O r$
(NS)	$\sigma_O r^2$
$(NS)(1234)$	$\sigma_O r^{-1}$
(13)	$\sigma_O \rho$
$(12)(34)$	$\sigma_O \rho r$
(24)	$\sigma_O \rho r^2$
$(14)(23)$	$\sigma_O \rho r^{-1}$

Solution 4.2

It is $\{e, \sigma r, r^2, \sigma, r, \sigma r^2\}$. Order 6.

Solution 4.3

It is
$$\rho(\sigma r)\rho^{-1} = (NS)(23)(NS)(123)(NS)(23) = (NS)(132) = \sigma r^{-1}.$$

We can also see this by observing that

$$\rho(\sigma r)\rho^{-1} = \sigma(\rho r \rho^{-1}) \quad \text{(because } \sigma \text{ and } \rho \text{ commute)}$$
$$= \sigma r^{-1} \quad \text{(from the property of the dihedral group } D_3\text{)}.$$

Solution 5.1

If n_p, n_q and n_r are all at least 3, then
$$\frac{1}{n_p} + \frac{1}{n_q} + \frac{1}{n_r} \leq 1,$$
which implies that
$$\frac{2}{|G|} \leq 0,$$
which is impossible.

Solution 5.2

If n_q and n_r are both at least 4, then
$$\frac{1}{n_q} + \frac{1}{n_r} \leq \tfrac{1}{2},$$
which implies that
$$\frac{2}{|G|} \leq 0,$$
which is impossible.

OBJECTIVES

After you have studied this unit, you should be able to:

(a) understand how to calculate the symmetry groups of the dihedron with n equatorial vertices ($n = 1, 2, \ldots$), tetrahedron, cube and dodecahedron;

(b) understand the duality between the cube and the octahedron and between the icosahedron and the dodecahedron;

(c) use the catalogue of finite isometry groups of $\mathbb{R}^3$;

(d) find the symmetry group of a simple regular solid figure;

(e) produce a simple solid figure which has a given group as its group of symmetries;

(f) perform computations involving the elements of finite three-dimensional symmetry groups.

INDEX

central inversion of the plane 11
cyclic group 8
DIH_n 19
dihedral group 8
dihedron 19

fixed point theorem 7
marked n-gons 11
orthogonal group in $\mathbb{R}^2$ 8
orthogonal group in $\mathbb{R}^3$ 13
orthonormal basis 14

Platonic solids 5
regular solids 5
special orthogonal group in $\mathbb{R}^2$ 10
special orthogonal group in $\mathbb{R}^3$ 14
symmetries of a solid figure F 19